Elastoplastic Behavior of Highly Ductile Materials

Maosheng Zheng · Zhifu Yin · Haipeng Teng ·
Jiaojiao Liu · Yi Wang

Elastoplastic Behavior of Highly Ductile Materials

Maosheng Zheng
Northwest University
Xi'an, Shaanxi, China

Haipeng Teng
Northwest University
Xi'an, Shaanxi, China

Yi Wang
Northwest University
Xi'an, Shaanxi, China

Zhifu Yin
Institute of Yanchang Petroleum
Group Co. Ltd.
Xi'an, Shaanxi, China

Jiaojiao Liu
Northwest University
Xi'an, Shaanxi, China

ISBN 978-981-15-0905-6   ISBN 978-981-15-0906-3   (eBook)
https://doi.org/10.1007/978-981-15-0906-3

This Springer imprint is published by the registered company Springer Nature Singapore Pte Ltd.
The registered company address is: 152 Beach Road, #21-01/04 Gateway East, Singapore 189721, Singapore

Preface

Nowadays, the industrialization significantly promotes the progress and application of elastoplasticity. Various new demands and phenomena emerged, which related to the development of materials with high ductility and strength in recent years. Therefore, it requires an appropriate description to present the wide classes of elastoplastic behaviors of such materials.

In 2008, a book entitled "Notch Strength and Notch Sensitivity of Materials" (authored by Zheng X., Wang H., Zheng M., and Wang F. H., Science Press, Beijing, China) was published, which is devoted to quantitatively characterize the notch strength and notch sensitivity of both ductile and brittle materials, and aimed to formulate the fracture criteria for notched structures. The present authors have made up their mind to publish the current book as an accompanying contribution to ductile materials, especially due to their wide applications.

In writing this book, the fundamental knowledge in the general elastoplastic theories was carefully selected from various formulations to date in the first three chapters, which are relevant to the study of the elastoplastic behaviors of highly ductile materials.

The first author has investigated elastoplastic behaviors of materials with high ductility since the early 1990s and has lectured materials and elastoplastic mechanics more than 20 years, various books and research papers both in Chinese and in English concerning this subject are piled at hand, and this book addresses the latest phenomena and formulations of elastoplastic behaviors of highly ductile materials in the last seven chapters comprehensively. Various approaches developed by the authors are included among the contents of this book in a certain room as well.

The main purpose of this book is to expedite the application of elastoplasticity theory to analyze engineering problems of elastoplastic behaviors of highly ductile materials in practice. It is our great pleasure if the readers including researchers, engineers, and students in the relevant fields could get valuable knowledge from this book.

As a foundation, the fundamental knowledge of general elastoplastic theories is introduced in Chaps. 1–3. Chapter 1 addresses the fundamental assumptions in elastoplastic mechanics; consequently, some models, assumptions, and relations are described in detail. Explanations for models and assumptions are to the extent that is sufficient to understand the subject of elastoplastic behaviors of ductile materials; Chap. 2 displays the description of physical relationship in elastoplastic mechanics, which includes the generalized Hooke's law, plastic yielding criteria in plastic mechanics, and relationship of incremental strain depending upon stress status in plastic mechanics. The damage evolution in ductile materials is presented quantitatively as well; Chap. 3 presents the solutions to some typical problems of elastoplasticity, such as the planar problems in elastic mechanics, elastoplastic analysis and plastic limit analysis of thick-walled cylinders, and elastoplastic bending and plastic limit analysis of beams; stress analysis of tube under uneven external load is provided, too.

Chapter 4 especially represents the transfer and relief of stress concentration at the notch root of structures of ductile materials, and analytical expressions are presented; Chap. 5 shows the analysis of elastoplastic deformation in the manufacturing process of bimetal composite pipes, which can be exceptionally considered as a typical application of the classical overmatching problem in engineering; Chap. 6 devotes to the plastic buckling of tube bending, as well as the strain hardening effect; Chap. 7 gives defect effect on pipe bending behavior, and both diffusive and localized corrosive defects are involved separately; Chap. 8 addresses thermal stress problems, and grain-reinforced Al matrix composite is particularly taken as a typical example due to its potential application; Chap. 9 describes the general description of fatigue problems first, then the uniform fatigue life equation for both low-cycle fatigue and high-cycle fatigue conditions and its improvement are peculiarly presented, and the mean stress effect is included. Chapter 10 provides energy absorption of highly ductile materials and characteristics of energy-absorbing components first, and then both horizontally compressed ring and axial compressions of round tube are given specially.

This book presents the research on some basic phenomena and laws of highly ductile materials during elastoplastic deformation, and their typical engineering applications, which are from the available literature and our group. The authors wish that the publication of this book would contribute to the relevant research in both theoretical research and engineering applications.

At this moment, the authors would like to express sincere thanks to relevant colleagues for their remarkable works in early days, especially Prof. Zheng X. L. (Zheng Xiulin) from Northwestern Polytechnic University in China for his crucial supervision and outstanding contributions in this field. The authors acknowledge their arduous and creative works.

It should be noted that the early research works on the ductile behaviors of metals were done by Zheng M. in Northwestern Polytechnic University, Xi'an Jiaotong University in China, Lappeenranta University of Technology in Finland, and Freiberg University of Mining and Technology in Germany. The authors also wish to express thanks to the relevant colleagues and institutions. In addition, Prof.

Sih G. C. of Lehigh University is acknowledged for his meaningful enlightenments and valuable discussions. The authors would like to express special gratitude to Prof. Niemi E. from Lappeenranta University of Technology in Finland for his kind hosting, Alexander von Humboldt Foundation of Germany for its support, and Prof. Kunz L. from Institute of Materials Physics of Czech Republic Science Academy for his effective collaboration.

The authors wish this work would make the contributions to relevant fields as a paving stone.

Xi'an, China

Maosheng Zheng
Zhifu Yin
Haipeng Teng
Jiaojiao Liu
Yi Wang

Contents

Chapter 1
Introduction

Abstract The fundamental assumptions of elastoplastic mechanics, task of elastoplastic mechanics, deformation characteristics of material, simplified constitutive model of deformed body and equilibrium equations, as well as limitation of some approaches are presented in this chapter.

1.1 Fundamental Assumptions of Elastoplastic Mechanics

Elastoplastic mechanics is a branch of solid mechanics. It is a discipline that studies the deformation law of elastic and elastoplastic objects. The derivation process is reasonable and rigorous, and the calculation result is accurate. It is the basis in analyzing and solving many engineering problems.

Elastic Mechanics

Since Hooker R. proposed the law that the deformation of an elastic medium is proportional to the external force (Hook's law) in 1678, the mathematical theory for elasticity is formulated by French scientists Navier C. L. M. H., Cauchy A. L., and de Saint-Venant A. J. C. B. in the 1820s. They gave the concepts of strain, strain component, stress, and stress component reasonably, and established the equilibrium equation, geometric equation, coordination equation, as well as the generalized Hooke's law for isotropic and anisotropic materials, which lays the theoretical foundation of elastic mechanics.

Plastic Mechanics

Beginning with the yielding condition of soil proposed by de Coulomb C. A. in 1773, the yielding condition of maximum shear stress was developed by Tresca H. in 1864. de Saint Venant A. J. C. B. believed that the direction of the maximum increment of plastic deformation should be consistent with that of the maximum shear stress during deformation. According to this idea, Levy M. extended the plastic stress–strain relationship to the three-dimensional case in 1871. Thereafter, von Mises R. put forward a plastic yielding condition by means of strain energy and independently

© Springer Nature Singapore Pte Ltd. 2019
M. Zheng et al., *Elastoplastic Behavior of Highly Ductile Materials*,
https://doi.org/10.1007/978-981-15-0906-3_1

presented the expression to correlate plastic strain increment and stress, which is the same as that is proposed by Levi. These all belong to the theory of rigid plasticity models since plastic strain increment is considered. Since then, Prandtl L. and Reuss A. advanced new expression to correlate the three-dimensional plastic strain increment and stress that includes the elastic strain increment. This is the incremental theory in plastic mechanics. At the same time, Hencky H., Nadai A. L., and Iliushin A. A. presented the deformation theory of plastic mechanics. In this period, small-scaled elastic–plastic mechanics is widely used to solve practical engineering problems. Many important concepts, rules, and principles have been established in theory, and many effective methods for solving problems have been given. In 1913, a German mechanics scientist von Mises R. brought forward an equivalent stress-based plastic yielding criterion for material. In 1924, another German mechanics scientist H. Hencky gave a meaningful interpretation to von Mises criterion. In Hencky's interpretation to von Mises criterion, he considered that if the elastically distortional energy density reaches a critical value the material element yields plastically.

1.1.1 Task of Elastoplastic Mechanics

With the developments of production in industry and agriculture, and other science and technology fields, it provides a lot of topics for elastoplastic mechanics, which promotes the progresses of elastoplastic mechanics, and also supplies more reliable technology in measurement methods and advanced calculation tools for the numerical assessment in elastoplastic mechanics.

At present, elastoplastic mechanics has been widely used in civil engineering, machinery, water conservancy, aviation, shipbuilding, nuclear energy, metallurgy, mining, materials, and other engineering fields. With the deepening of research work, elastoplastic mechanics will play an increasingly important role in various projects.

The establishment of the basic equation of elastoplastic mechanics needs to be studied from three aspects: geometry, kinematics, and physics. In kinematics, it is mainly to establish the equilibrium condition of the object, not only the object as a whole should be balanced, but also any part of the object (material unit) should be in equilibrium. There are two types of mathematical equations that reflect this law, namely the motion (or equilibrium) differential equation and the boundary conditions of the load. Both of the above equations are independent of the mechanical properties of the material and belong to the universal equation.

In physics, the relationship between stress and strain or increment of stress and increment of strain is established. This kind of correlation is often called constitutive relation, and it describes the mechanical properties of materials in different environments. In elastoplastic mechanics, the study of constitutive relations is very important. Since the characteristics of matter in nature are diverse and the working environment in which they are located is different, the study of constitutive relationship of matter is a complex and fundamental work. Since objects are continuous, each adjacent small unit is interconnected during deformation, and by studying the relationship between displacement and strain, the coordination condition of the deformation can be obtained. There are two types of mathematical expressions that reflect the continuous law of deformation, namely geometric equations and displacement boundary conditions.

1.1.2   Basic Assumptions of Elastoplastic Mechanics

In the analysis of elastoplastic mechanics, the following simplifying conditions are often used.

(1) The object is continuous, and its stress, strain, and displacement can be described by a continuous function.
(2) The object is uniform and isotropic, each part has the same properties, and the physical constant does not change with position and direction.
(3) The deformation is very small, and the displacement of each point in the object after deformation is much smaller than the original size of the object itself, so that the geometric change caused by the deformation can be ignored.

In fact, the basic theory comes from practice and summarizes the law in practice. The law is often complicated, and the corresponding theory must be established through basic assumptions. The following shows more details of general assumptions and rules in elastoplastic mechanics:

(1) *Uniform continuous hypothesis*, the medium is uniformly continuous without gaps in the entire object. Although the actual material consists of discontinuous particles microscopically, the scale of the inhomogeneity is much smaller than the macroscopic observation range.
(2) *Elastic properties of the material are not affected by plastic deformation*. When any point in the object is in a plastic state, the deformation is divided into elasticity and plasticity. The elastic strain is always linear with the stress regardless of the amount of plasticity, and the loading and unloading processes obey the generalized Hooke's law.
(3) *Ignoring the influence of time on the properties of the material*, it is assumed that the stress and strain in the object are only related to the magnitude of the applied load, regardless of the time. It is assumed that the deformation rate, strain rate, and other indicators are the increments of displacement and strain, while it ignores the applied time period. The force and deformation after loading are fixed values and do not change with time.
(4) *Considering the stable material and the increase of load gradually only*. The stable material refers to the product of the change of the stress and the change of the strain greater than or equal to 0 in any single period of loading. The theory of constitutive relations is established under this assumption.

(5) The deformation of the object is in *a small deformation range*, regardless of the dimensional change caused by the deformation.
(6) If the object is in an *unstressed natural state*, the *stress at each point is zero.*
(7) *Isotropic* means that the nature of the material along all directions is the same.
(8) The *elastic constitutive relation is linear*, and the *plastic constitutive relation is nonlinear.*

1.1.3   Solving the Problem of Elastoplastic Mechanics

When solving an elastoplastic problem, it is necessary to give the shape of the object, the constitutive relation, and physical constant of the material for each part of the object, indicating the load on the object and the connection with other objects, i.e., the boundary condition. For kinetic problems, initial conditions are also given.

The mathematical method for solving the elastoplastic problem is to get the functions of displacement, strain, and stress according to the geometric equation, the physical equation, the equilibrium equation, the boundary conditions, and initial conditions of the force and displacement. It is very effective to solve some simple problems in this way.

There are two types of methods in solving elastoplastic mechanics problems:

(1) *Approach of exact solution*, i.e., the solution satisfies all equations of elastoplastic mechanics;
(2) *An approximate solution*, i.e., it is based on the nature of the problem, and reasonable simplifying hypothesis is used to obtain an approximation.

The finite element method has been continuously developed with the progress of computer technology, which provides extremely favorable conditions for the development of elastoplastic mechanics. In general, there is no limitation for geometries of components and objects, and it can produce correct results for a variety of complex geometries and physical relationships.

The purpose of solving the elastoplastic problem is to find the proper stress and displacement of each point in the object, i.e., the stress field and the displacement field. In fact, it means to give the stress field and displacement field that generated in the object to response the external actions (including temperature, external force, etc.) acting on the object.

The basic equations of elastoplastic mechanics always govern the general law of the relationship among stress, strain, and displacement inside the object. The boundary conditions specifically give the particular constraints of problem on each boundary, and each specific problem is reflected in the respective boundary conditions. Then, the boundary conditions and the basic equations of elastoplastic mechanics together constitute the complete formulation of the elastoplastic mechanical problem.

According to different types of boundary conditions, the boundary value problem is often divided into the following three categories.

The first type of boundary value problem is to get the equilibrium stress and displacement fields under the condition of the body force and surface force on boundary of a object given; that is, the so-called *stress boundary is known*.

The second type of boundary value problem is to solve the equilibrium stress and displacement fields under the condition of the body force of a given object and the displacement on the boundary of the object given; that is, the so-called *displacement boundary is known*.

The third type of boundary value problem is a so-called *hybrid boundary value problem*, on a part of the boundary the surface force is known, and on the rest the displacement is given.

When solving the above boundary value problems, there are three different processing methods, namely

(1) *Displacement method*, using displacement as the basic unknown function to solve the boundary value problem. At this time, all the unknown quantities and the basic equations are converted into displacements. Usually, it applies the problem when the boundary condition belongs to displacement boundary condition (the second type of boundary value problem).

(2) *Stress method*, using stress as the basic position function to solve the boundary value problem. At this time, all the unknown quantities and the basic equations are converted into stresses. Usually, it applies the problem when the boundary condition belongs to stress boundary condition (the first boundary value problem).

(3) *The hybrid method*, it corresponds to the third type of boundary value problem, a part of the displacement component and the rest of the stress component should be used as the basic unknown quantity, and a set of hybrid solutions will be obtained.

As to the pure elastic problem, there exist the following features of the solutions:

(1) *Superposition Principle*

In the range of linear elastic mechanics, the relationships among stress, strain, and displacement are all linear due to small deformation of the object. The so-called principle of *superposition principle* means that the result of sum of two sets of external forces (bulk force and surface force) acting on the object is equal to the sum of the effects of the two external forces, respectively, generated inside the object separately.

It is well known that the internal effect of an object is caused by an external force. If the surface force or volume force of an object is known, the resulting stress and displacement must satisfy the basic equations and boundary conditions of the elastic theory.

(2) *Uniqueness Theorem of Solution*

The uniqueness of the solution can be proved. It assumes that the object is in a natural state; that is, there is no initial stress; the displacement is considered to be small and does not affect the external force, so the *superposition principle* can be applied.

Under the above premise and certain boundary conditions (surface force and displacement, as well as certain volume force), the solution of its internal stress and strain components within the object is unique, which is called the uniqueness theorem of the solution.

However, for the general elastoplastic problem, the stress and deformation must be solved step by step with the application history of the load in principle.

In the actual process, the elastoplastic problem can be divided into three categories: elastic phase, elastoplastic phase and plastic limit state according to the condition of its load. However, due to mathematical difficulties, many problems are not solved accurately, and an approximate method is needed to solve them. Many practical problems in engineering are that only the ultimate load is required. For this reason, the approximate solution is obtained by using the limit theorem according to the plastic limit state without considering the loading history. This method of analysis is called *limit analysis*.

When the structure reaches the plastic limit state, its elastic strain is much smaller than the plastic deformation and can be neglected. Therefore, it can be assumed that the material is ideal elastic–plastic one, in which case the structure may enter the plastic state in whole or part. Since the plastic deformation can be freely developed in the region in the plastic stage, the entire structure can no longer withstand a larger load and become a plastic deformation mechanism (or a limit mechanism). The limit of this load is called the *plastic limit load*, which is called the ultimate load. The plastic limit theorem, or the upper and lower limit theorems, can be used to calculate the limit value.

Only the true displacement field, strain field, and stress field are both static and maneuverable.

The principle of virtual work can be expressed as follows: Under the action of external force, a deformable solid in equilibrium, if a small virtual deformation is given, the virtual work done by the external force on the virtual displacement is equal to the virtual work done by the internal force (stress).

The principle of maximum plastic work means that the true plastic work is the largest, indicating that the material always has to produce the maximum energy consumption when it undergoes plastic deformation.

As to solving the boundary value problem of plastic mechanics, the solution is often difficult due to the nonlinear nature of the constitutive equations, and therefore many powerful approaches have been developed. The most widely used ones are (1) solving the static problem and (2) the upper and lower limit method. These two methods have been commonly used in plastic mechanics. Extensively the upper and lower limit method, it could derive many important analytical results in structural plastic limit analysis and stability analysis as well as in plastic forming of metals.

The slip-line method, principal stress method, parameter method, weighted residual method, and energy method are also very important analytical methods, but it is convenient under certain conditions only.

The descriptions of above two common methods are given as follows:

(1) *Static Problem*

As a solution to the simple problem, the characteristics are that the number of the equilibrium equation, the yielding condition, and other constraints are equal to the number of the unknown quantities, so that the unknown quantities can be found without using the nonlinear constitutive equation in elastoplastic mechanics. Most of the one-dimensional problems in plastic mechanics fall into this category, for example, rotating disks, thick-walled cylinders, solid and hollow torsion round shafts, and elastoplastic bending of beams with various cross sections. When solving, the ideal elastoplastic mechanical model is generally used. Although this kind of problem is simple and convenient to solve, it is often encountered in engineering practice, and it has great application value.

(2) *Upper and Lower Limit Method*

It is a very useful analytical method. In general, it is very difficult to find an exact solution that satisfies all requirements in plastic mechanics, so it is meaningful to find solutions that satisfy some of the equations and to give a proper estimation of the properties of these solutions. The plastic mechanics equations fall into two categories.

The first type includes the equilibrium equation, the plastic yielding equation, and the boundary conditions of the force, which is called static condition. The geometric requirements are not included in these conditions at all. If a solution satisfies the above static condition, it becomes a static solution, which gives the stress with values of no more than that of the complete solution (the solution satisfying all the conditions of plastic mechanics).

In another type of equation, the work done by the external force is equal to the condition of the internal dissipated work and the geometric boundary condition of the structure is met. The static force is not considered. This method is called the maneuver method. In general, the result of stress in maneuver method is not less than the complete solution. So, it is also known as the upper limit method because the minimum load may be equal to the ultimate load obtained by the complete solution. The upper limit method has been widely used in plastic forming problem of metals and plastic limit analysis of plate and shell. In the upper limit method, the upper limit value can always be found according to the principle of virtual work in mechanics and a certain failure mechanism. While the destruction mechanism can be found through experimental methods, the most reasonable failure mode is consistent with the experimental results.

The newly developed upper limit elementary technology in recent years and the upper limit method for dealing with the continuous velocity field problem with failure mode are with features clear in concept and convenient in use.

Generally, a total amount of 15 unknown functions are included in the problem of elastic mechanics, which will be solved by 15 equations. For an isotropic elastic medium, there are three equilibrium differential equations, six geometric equations (also differential equations), and six physical equations (generalized Hooke's law). The number of equations equals to the number of unknown functions.

In each specific problem, the boundary conditions on the surface of the elastic medium should also be given as a supplementary condition of these equations to form a closed fixed solution problem.

1.2 Deformation Characteristics of the Material

1.2.1 Stress–Strain Curve

Although different materials have their specific tensile curves, they also have some common rules in general. When the deformation is quite small, that is, if the stress is less than the elastic proportional limit (see Fig. 1.1), the relationship between stress and strain is linear and thus deformation can be recovery; i.e., after the external load is removed, the object can be completely restored to its initial state before the deformation, and there is no residual deformation and residual stress in the object.

Let the initial cross-sectional area of the test piece be A_0, and P is the external load. Assuming the distribution of stress is uniform, if the stress in the test piece is σ, then

$$\sigma = P/A_0, \tag{1.1}$$

If the gauge length of the specimen is l_0, and the elongation of the specimen is Δl, then the nominal strain is

$$\varepsilon = \Delta l/l_0. \tag{1.2}$$

Fig. 1.1 Stress–strain curve for tensile process of a smooth cylinder bar

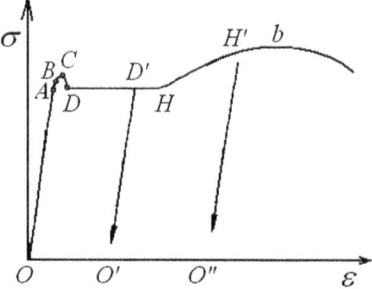

In Fig. 1.1, *OA* is the proportional deformation stage, and relationship between stress and strain is linear, which can be described by Hooke's law,

$$\sigma = E\varepsilon, \tag{1.3}$$

in which, E is elastic modulus of material, and it is a constant in elastic mechanics.

The stress corresponding to point *A* is the proportional limit. The segment from *A* to *B* cannot be expressed by a linear relationship in principle though the deformation is still elastic, and the stress corresponding to point *B* is called the elastic limit. For many materials, the distance from point *A* to point *B* is very small. The stresses corresponding to points *C* and *D* are referred to as the upper yielding limit and the lower yielding limit of the material, respectively. When the stress reaches point *C*, the material begins to form plastic yielding. Generally speaking, the upper yielding limit is greatly affected by external factors, such as the shape and size of the specimen, and the loading rate has an effect on it as well. Therefore, in practical applications, the lower yielding limit is generally used as the yielding strength of the material, which is denoted as σ_s.

The stress from point *C* to point *D* exhibits abrupt drop, but the strain is continuously increasing, while from point *D* to point *H* there is a line close to the horizontal. In this stage, although the stress is not increased, the strain is continuously increasing, so it is called the plastic flow stage. The plastic flow stage is rather long for some material. The strengthening phenomenon starts from point *H*; that is, the strain can be increased only if the stress increases. If unloading in the plastic yielding or strengthening stages of material, the unloading process is along the line *D'O'* or *H'O''*; see Fig. 1.1. It can be seen from Fig. 1.1 that *D'O'* and *H'O''* are parallel to the *AO* line; i.e., though the material is plastically deformed, the elastic properties of the material do not change. If reloading begins from *O''*, the loading process continues along the *O''H'* line until the material begins to yield after *H'*. Before the point *b*, the specimen is in a uniform strain state. After reaching to the point *b*, the specimen tends to have a necking phenomenon. If the tensile process is continued, the deformation will be concentrated in the necking region, and the cross area there is reduced rapidly, so the specimen will be broken there soon.

Stress–Strain Curve with no Obvious Plastic Yielding Point

As shown in Fig. 1.2, the stress–strain curve has no obvious plastic yielding point for some materials. Thus, the plastic yielding limit is specified by the stress corresponding to 0.2% plastic strain in this case, and it is recorded as $\sigma_{0.2}$.

While in Fig. 1.3, there is another case, the unloading (compression) is continued from point *O''*, the stress σ_s'' corresponding to the yield point *C* at the reversal loading is smaller than σ_s' and the initial yield limit σ_s. Here, σ_s' is the yielding limit from the point *O'* on the stress–strain curve in tensile process. When the plastic deformation increases, it exhibits the increase of yielding strength in one direction but decreases in the opposite direction. This phenomenon was first discovered by the German scientist Bauschinger J., which is called the Bauschinger effect. It is

Fig. 1.2 Stress–strain curve
with no obvious yield point

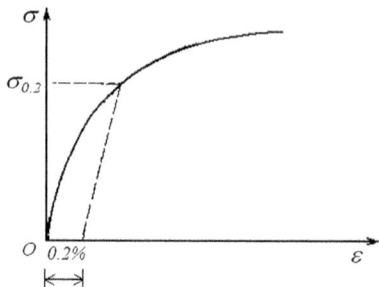

Fig. 1.3 Bauschinger effect

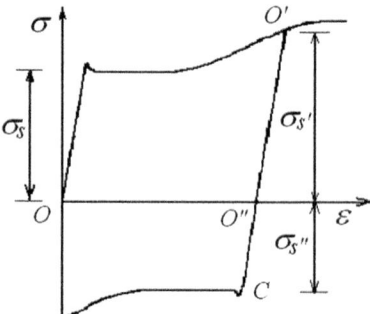

generally believed that this phenomenon is caused by residual stress between the
grain boundaries of the polycrystalline material. The Bauschinger effect gives the
material an anisotropic property. If the increasing value of the yielding strength in
one direction is equal to the decrease of the yielding strength in the opposite
direction, it is called the ideal Bauschinger effect. Since the mathematical
description of this effect is more complicated, it is generally simplified in plastic
mechanics.

Generally, in tensile test, σ is not the true stress because the actual cross section
of the specimen is gradually reduced during the tensile process, and σ is defined as
the tensile force divided by the initial cross-sectional area. This issue has little effect
on the accuracy of the stress–strain curve before point b of the stress–strain curve in
Fig. 1.1. But after the point b, the cross section of the specimen shrinks intensively
and locally, and the large change of the cross-sectional area will have a significant
effect on the evaluation of the stress.

If the true stress on the cross section of the specimen is represented by σ_T, and
A is the actual cross section of the specimen at current moment before point b of the
stress–strain curve in Fig. 1.1, then there is

$$\sigma_T = P/A. \tag{1.4}$$

Since A is smaller than the initial cross-sectional area A_0, there should be $\sigma_T > \sigma$, and σ is called nominal stress. According to the volume incompressible assumption, before point b of the stress–strain curve in Fig. 1.1 there exists

$$A_0 \cdot l_0 = A \cdot l, \quad A = A_0 \cdot l_0/l \tag{1.5}$$

where l_0 and A_0 are the initial length and initial cross-sectional area of the specimen, and l and A stand for the current length and cross-sectional area of the specimen.

Substituting Eq. (1.5) into Eq. (1.4), it yields

$$\sigma_T = P/A = \sigma(1 + \varepsilon). \tag{1.6}$$

According to Eq. (1.6), the true stress–strain curve is easily obtained from the nominal stress–strain curve before point b of the stress–strain curve in Fig. 1.1. For rather large plastic deformation, the true strain is also defined by

$$e = \int_{l_0}^{l} \mathrm{d}l/l = \ln(l/l_0) = \ln(1 + \varepsilon). \tag{1.7}$$

However, after necking, i.e., after point b of the stress–strain curve in Fig. 1.1, the true stress and strain could not be expressed by Eqs. (1.6) and (1.7) though the tensile bar is an initial smoothed one.

Bridgeman Correction

In fact, necking is equivalent to notch in the tensile specimen. Although prior to necking, the stress state in the specimen is uniaxial, whereas after necking the stress state becomes triaxial one in nature, as shown in Fig. 1.4; i.e., radial and tangential

Fig. 1.4 Stress states in necking region

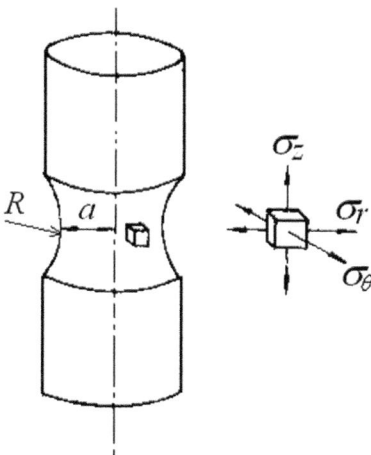

stresses appear in the volume element in the necked part. The longitudinal stress
which promotes the specimen undergoing further plastic deformation is increased.

Bridgeman corrected the average stress at the necking part according to the
following assumptions [1]: (1) Assume that the shape of the neck is an arc with a
curvature radius of R; (2) the cross section of the neck part is still maintained
circular shape throughout the test with a radius of a; (3) in the cross section of the
necking zone, the strain is constant and independent of the distance r from the
centerline of the cross section; (4) the von Mises yielding criterion is applicable.

According to Bridgeman's analysis, the relationship between the axial true stress
σ^* and the average true stress is expressed by Eq. (1.8) [1],

$$\sigma^* = \frac{\sigma_T}{(1 + 2R/a) \cdot [\ln(1 + a/(2R))]}. \tag{1.8}$$

Simultaneously, an empirical relationship between a/R and ε was well established [2, 3],

$$a/R = 1.1(\varepsilon_a - \varepsilon_u). \tag{1.9}$$

In Eq. (1.9), ε_a is the current axial strain $\varepsilon_a = \ln(A_0/A)$ and ε_u is uniform elongation of the material.

These equations are with high accuracy in most cases. The corrected true stress–
strain curve is shown in Fig. 1.5.

The actual stress distributions at the neck cross section are as follows according
to Bridgeman [2, 3]

$$\sigma_\theta = \sigma_r = \frac{\sigma_{am}}{(1 + 2R/a)} \left[\frac{\ln(\frac{a^2 + 2aR - r^2}{2aR})}{\ln(1 + a/2R)} \right], \sigma_z = \frac{\sigma_{am}}{(1 + 2R/a)} \left[\frac{1 + \ln(\frac{a^2 + 2aR - r^2}{2aR})}{\ln(1 + a/2R)} \right] \tag{1.8'}$$

where r is the distance from the centerline of the cross section; σ_{am}, σ_z, σ_r, and σ_θ
represent the average axial true stress, axial, radial, and hoop stresses, respectively.

Fig. 1.5 Stress–strain curve
of ductile metal

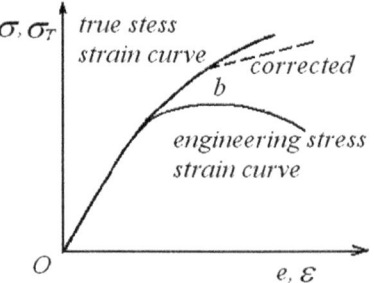

1.2.2 Basic Models of Materials

Different deformation models can be used for different materials in varying application fields. When determining the mechanical model, it is important to pay special attention to the fact that the selected mechanical model must conform to the actual conditions of the material, because this is the only way to make the calculation reflect the actual stress–strain state in the structure or component. On the other hand, it should be noted that the mathematical expression of the selected mechanical model should be simple enough so as to make the specific mathematical problem solvable.

For some materials, there is a linear relationship between stress and strain in the elastic deformation phase (shown in Fig. 1.6). At this stage, the stress, strain, and displacement caused by the external load are independent of the loading order and history. After the external load is removed, the object is completely restored to the initial state, and there are no residual stress and residual deformation in the object. At this time, the mathematical expression of the relationship between stress and strain is represented by Eq. (1.3). For some materials, even if the stress–strain relationship in the elastic phase is exactly not linear, Eq. (1.3) is an approximate expression. However, the stress does not increase in plastic yielding stage, see Fig. 1.6. *This is the ideal elastic–plastic model.*

When considering strain hardening properties of material, a *linear strain hardening model* for the elastoplastic material can be used (see Fig. 1.7). There are two lines in Fig. 1.7, namely OA and AB lines, and the analytical expressions are

$$\sigma = E\varepsilon \quad (\varepsilon < \varepsilon_s), \quad \sigma = \sigma_s + E_1(\varepsilon - \varepsilon_s) \quad (\varepsilon > \varepsilon_s) \tag{1.10}$$

In Eq. (1.10), E and E_1 are the slopes of the line segments OA and AB, respectively. Material with such a stress–strain relationship is called an elastoplastic linear strengthening material. Since OA and AB are two straight lines, they are sometimes called *bilinear strengthening models.*

For some materials, if the slope of AB is small enough, it does not cause a large error as considering it as an ideal elastoplastic model, thus the analysis and calculation can be greatly simplified, and this approximate mechanical model is

Fig. 1.6 Ideal elastic–plastic model

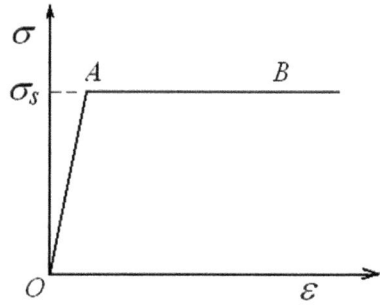

Fig. 1.7 Linear strain
hardening model

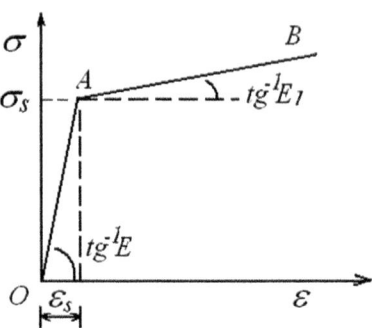

accurate enough. When slope of *AB* is too large to be neglected, it should be
calculated according to Eq. (1.10). Although difference between the ideal elastoplastic model and the linear hardening model is not much, the analysis and calculation involved in the latter case are much more complicated.

Another strain hardening model is *power hardening one*, which can be used
sometimes as well, that is,

$$\sigma = K\varepsilon^n, \tag{1.11}$$

where n is exponent of the strain power hardening, which is between 0 and 1, and
K is the strength coefficient. See Fig. 1.8. Equation (1.11) is also called Hollomon
formula [1]. If $n = 0$, it becomes ideal elastic–plastic model. The greater the value
of n, the higher the resistance of the material to resist the further plastic deformation. The index n of most metallic materials is between 0.05 and 0.50 (see
Table 1.1) [1].

Experiments have shown that the higher the strength of the material, the lower
the value of n. In general, the n-value of the face-centered cubic metal is higher than
that of the body-centered cube [1].

In many practical engineering problems, the elastic strain is much smaller than
the plastic strain, so the elastic strain can be ignored. If the strain hardening effect is
not considered, the model is called a *rigid plasticity model*; see Fig. 1.9. In this

Fig. 1.8 Power hardening
model

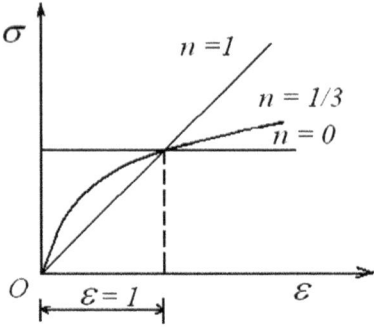

Table 1.1 n-value for some metallic materials

Metals	Austenitic stainless steel	Copper	Brass	Steel	Iron	Aluminum
n	0.45–0.55	0.3–0.35	0.35–0.4	0.15–0.25	0.05–0.15	0.15–0.25

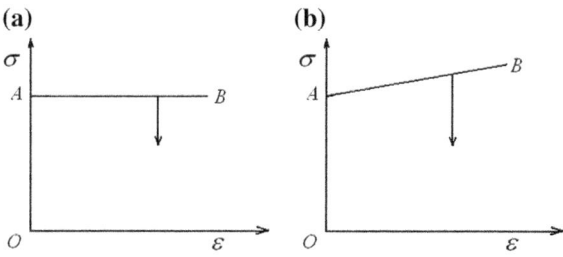

Fig. 1.9 Rigid plasticity model

case, the assumption is: The stress should be zero before it reaches the yielding limit. In Fig. 1.9a, the line segment AB is parallel to the ε-axis. The unloading line is parallel to the σ-axis. Figure 1.9b shows a rigid plasticity model with linear train hardening properties, and the unloading line is also parallel to the σ-axis.

In plastic mechanics, the rigid plastic mechanics model is of great significance. In most cases of plastic forming theory, the plastic strain is generally much larger than the elastic strain. It is reasonable to ignore the elastic strain and consider the plastic strain only. It has little effect on the overall result. The rigid plastic model is used easily to perform the calculation in mathematics. Such a greater simplification enables many complex problems get analytical expressions. Among the several mechanical models mentioned above, the ideal elastoplastic model, the power strengthening model, and the rigid plastic mechanics model are the most widely used ones.

1.3   Constitutive Model of the Deformed Body

1.3.1   Simplified Model of Uniaxial Stress–Strain Relationship

The mathematical expression of the uniaxial stress–strain relationship can be obtained by fitting the experimental curve of uniaxial tension (compression) mathematically. However, in some practical engineering elastoplastic problem, such expressions are complex in analysis and calculation. Therefore, it is necessary to simplify expressions appropriately according to the actual characters of different materials so as to keep the main features of the material properties and make it convenient in mathematical treatment. The most commonly used models are as follows.

1. *Ideal Elastoplastic Model*

For low carbon steel or materials with low hardening rate, the hardening effect can be neglected as the strain is not too large, and the uniaxial tensile stress–strain curve is simplified as shown in Fig. 1.6. Once the stress reaches the plastic yielding limit σ_s, it can no longer be increased; if the material is unloaded, $d\sigma < 0$, it progresses along a line paralleling to line AO. The stress–strain relationship can be expressed as

$$\varepsilon = \sigma/E \ (\sigma < \sigma_s), \quad \varepsilon = \sigma_s/E + \lambda \ (\sigma = \sigma_s), \tag{1.12}$$

wherein λ is an undetermined quantity.

2. *Linear Hardening Model*

The continuous stress–strain relationship curve is approximated by two straight lines. As shown in Fig. 1.7, the first straight line represents the linear elastic deformation property, and its slope is E. The second straight line represents the hardening property, and the slope is represented by E_1. In general, E_1 is much smaller than E. At this time, the stress–strain relationship can be expressed as

$$\varepsilon = \sigma/E \ (\sigma < \sigma_s), \quad \varepsilon = \sigma_s/E + (\sigma - \sigma_s)/E_1 \ (\sigma > \sigma_s), \tag{1.13}$$

If the material is unloaded, $d\sigma < 0$, it progresses along a line paralleling to line AO.

3. *Power Exponential Hardening Model*

For most materials, the hardening curve is nonlinear. If a simple power function is used for the simulation, the stress–strain relationship can be expressed as

$$\sigma = E\varepsilon \ (\sigma < \sigma_s), \quad \sigma = K\varepsilon^n \ (\sigma > \sigma_s), \tag{1.14}$$

where K and n are material constants, which can be obtained by fitting the experimental curve. These two constants are not independent, and it requires the continuity at $\sigma = \sigma_s$,

$$\sigma_s = K(\sigma_s/E)^n. \tag{1.15}$$

If the material is unloaded, $d\sigma < 0$, it progresses along a line paralleling to initial tangential line of the curve around point O.

4. *Other Hardening Models*

In an actual loading process, the loading surface may change continuously with the deformation due to loading. Moreover, the change of the loading surface should satisfy the updated stress state. The point is always on the updated loading surface. The loading surface satisfying such condition is theoretically infinite. Therefore, the actual hardening law is very complicated, and it is not easy to completely determine loading surface by experimental methods. For the initially isotropic metal material, during the plastic deformation process, the material may have a preferred orientation of the crystal lattice, which leads to anisotropic (deformation-induced anisotropy) due to the plastic deformation. The anisotropic effect is more pronounced with deformation, which makes the evolution of the loading surface more complicated. In order to facilitate the practical application, the evolution of the loading surface has to be simplified. The typical simplified hardening models include: isotropic hardening, follow-up hardening model, mixed hardening model, anisotropic linear elastomer, and transversely isotropic material [4].

1.3.2 Limitations of Conventional Elastoplasticity Model

The conventional elastoplasticity models described in the above sections are based on their typical assumptions; therefore, it gives different stress–strain curves. These stress–strain curves may behave differently as compared to the actual loading response of material; the acceptability of accuracy of each model is upon the actual engineering requirement for one to judge; furthermore, in the viewpoint of elastic mechanics, everything is recoverable in elastic loading range; i.e., there will be no damage if the deformation is repeated in an elastic range for the cyclic loading of stress below the yield stress. However, in real materials, some local plastic deformation is accumulated for stress cycles less than the plastic yielding strength, which leads to the failure called fatigue. Therefore, the conventional plasticity has its limitations in the application to the mechanical design of machines and structures in engineering practice.

1.4 Equilibrium Equations

According to the basic assumption of elastoplastic mechanics, the object is continuous, its stress, strain, and displacement can be described by continuous functions inside the body, and they are uniquely determined from the applied loadings. At the same time, bodies are in equilibrium state, and the applied loadings could satisfy the equations of static equilibrium; the summation of forces and moments could be zero [4].

The entire body and then all parts must also be in equilibrium. Therefore, the entire body can be partitioned into appropriate sub-domains and the equilibrium principle could be applied to each region. Following this idea, equilibrium equations can be developed.

Take a volume element within a body as an example, which is in equilibrium. The stress components are shown in Fig. 1.10. As to static equilibrium, the forces

Fig. 1.10 Stress components
in a volume element

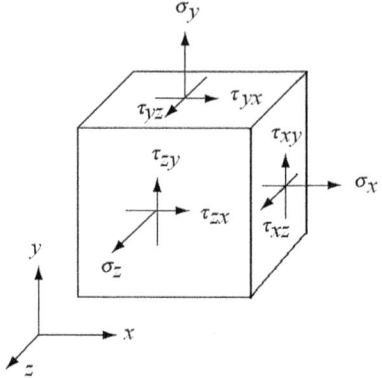

acting on this region must be balanced and thus the resultant force must vanish.
Writing this concept mathematically, it yields the following expressions

$$\frac{\partial \sigma_x}{\partial x} + \frac{\partial \tau_{yx}}{\partial y} + \frac{\partial \tau_{zx}}{\partial z} + F_x = 0$$

$$\frac{\partial \tau_{xy}}{\partial x} + \frac{\partial \sigma_y}{\partial y} + \frac{\partial \tau_{zy}}{\partial z} + F_y = 0 \qquad (1.16)$$

$$\frac{\partial \tau_{xz}}{\partial x} + \frac{\partial \tau_{yz}}{\partial y} + \frac{\partial \sigma_z}{\partial z} + F_z = 0$$

Or in short,

$$\sigma_{ji,j} + F_i = 0. \qquad (1.17)$$

In Eqs. (1.16) and (1.17), σ_{ji}, F_i, and $\sigma_{ji,j}$ express the components of stress, body
force, and differential stress component, respectively. It adopts the convention that
if a subscript appears twice in the same term for convenience, then summation over
that subscript from one to three is implied, for example,

$$a_{ii} = \sum_{i=1}^{3} a_{ii} = a_{11} + a_{22} + a_{33}$$

$$\sigma_{ji,j} = \sum_{j=1}^{3} \sigma_{ji,j} = \sigma_{1i,1} + \sigma_{2i,2} + \sigma_{3i,3} \qquad (1.18)$$

Stress can be written as matrix [4],

$$\text{stress matrix} = [\sigma] = \begin{bmatrix} \sigma_x & \tau_{xy} & \tau_{xz} \\ \tau_{yx} & \sigma_y & \tau_{yz} \\ \tau_{zx} & \tau_{zy} & \sigma_z \end{bmatrix} \qquad (1.19)$$

There are nine components in stress matrix, which are called the stress components, with σ_x, σ_y, σ_z referred to as normal stresses and τ_{xy}, τ_{yx}, τ_{yz}, τ_{zy}, τ_{zx}, τ_{xz} as the shearing stresses.

The symmetric characteristic property of stress components indicates

$$\sigma_{ij} = \sigma_{ji}, \text{ or } \tau_{ij} = \tau_{ji}. \tag{1.20}$$

Strain can be written as matrix,

$$\text{strain matrix} = [\varepsilon] = \begin{bmatrix} \varepsilon_x & \gamma_{xy} & \gamma_{xz} \\ \gamma_{yx} & \varepsilon_y & \gamma_{yz} \\ \gamma_{zx} & \gamma_{zy} & \varepsilon_z \end{bmatrix} \tag{1.21}$$

The symmetric characteristic property of strain components is similar to Eq. (1.20).

By using the more compact tensor notation, strain components are written as

$$\varepsilon_{ij} = (u_{i,j} + u_{j,i})/2. \tag{1.22}$$

Here, u_i is the displacement in i direction.

The Saint-Venant compatibility equations are

$$\frac{\partial}{\partial x}\left(\frac{\partial \gamma_{xy}}{\partial z} + \frac{\partial \gamma_{zx}}{\partial y} - \frac{\partial \gamma_{yz}}{\partial x}\right) = 2\frac{\partial^2 \varepsilon_x}{\partial y \partial z}$$

$$\frac{\partial}{\partial z}\left(\frac{\partial \gamma_{yz}}{\partial x} + \frac{\partial \gamma_{xz}}{\partial y} - \frac{\partial \gamma_{xy}}{\partial z}\right) = 2\frac{\partial^2 \varepsilon_z}{\partial x \partial y} \tag{1.23}$$

$$\frac{\partial}{\partial y}\left(\frac{\partial \gamma_{xy}}{\partial z} + \frac{\partial \gamma_{yz}}{\partial x} - \frac{\partial \gamma_{zx}}{\partial y}\right) = 2\frac{\partial^2 \varepsilon_y}{\partial x \partial z}$$

1.5   Principal Stresses

As stress and strain can be written in matrix forms, while the matrix has its principal value mathematically, therefore stress and strain have their principal values, i.e., the so-called *principal stresses* and *principal strains*, respectively. The principal values of stress can be obtained by the mathematic method of matrix principal value problem. Thus,

$$\det[\sigma_{ij} - \sigma\delta_{ij}] = -\sigma^3 + I_1\sigma^2 - I_2\sigma + I_3 = 0, \tag{1.24}$$

where σ represents the principal stresses; the basic invariants of the stress tensor can be expressed in terms of the three principal stresses σ_1, σ_2, and σ_3 as

$$I_1 = \sigma_1 + \sigma_2 + \sigma_3$$
$$I_2 = \sigma_1\sigma_2 + \sigma_2\sigma_3 + \sigma_3\sigma_1 \qquad (1.25)$$
$$I_3 = \sigma_1\sigma_2\sigma_3$$

where, I_1, I_2, and I_3 are called the *first, second,* and *third stress invariants,* respectively.

In the principal coordinate system, the stress matrix takes the special diagonal form,

$$\sigma_{ij} = \begin{bmatrix} \sigma_1 & 0 & 0 \\ 0 & \sigma_2 & 0 \\ 0 & 0 & \sigma_3 \end{bmatrix} \qquad (1.26)$$

From Eq. (1.26), it can be seen that in the principal coordinate system, all the shear stresses vanish and the state contains normal stresses only. This issue is also valid for the analysis of strain tensor.

Another important object is to investigate the transformation of stress and traction in the framework of principal stresses. Consider the general traction vector T_n that acts on an arbitrary surface as shown in Fig. 1.11. It is now to determine the normal and shear components of the traction vector, N and S. The normal component is simply the traction's projection in the direction of the unit normal vector $\boldsymbol{n}$, while the shear component is found by Pythagorean theorem,

$$N = T^n \times \boldsymbol{n}, S = (|T^n|^2 - N^2)^{1/2} \qquad (1.27)$$

By detailed derivation, it yields [4, 5]

$$N = \sigma_1 n_1^2 + \sigma_2 n_2^2 + \sigma_3 n_3^2, S^2 = \sigma_1^2 n_1^2 + \sigma_2^2 n_2^2 + \sigma_3^2 n_3^2 - N^2. \qquad (1.28)$$

Fig. 1.11 Schematics of principal stress

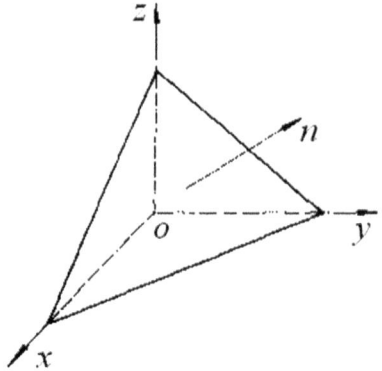

wherein n_1, n_2, and n_3 represent the cosines of intersection angles between n and the coordinate axes, respectively.

Additionally, the condition for vector n has to be unit magnitude

$$n_1^2 + n_2^2 + n_3^2 = 1. \tag{1.29}$$

It further obtains [5]

$$
\begin{aligned}
n_1^2 &= \frac{S^2 + (N - \sigma_2)(N - \sigma_3)}{(\sigma_1 - \sigma_2)(\sigma_1 - \sigma_3)} \\
n_2^2 &= \frac{S^2 + (N - \sigma_3)(N - \sigma_1)}{(\sigma_2 - \sigma_3)(\sigma_2 - \sigma_1)} \\
n_3^2 &= \frac{S^2 + (N - \sigma_1)(N - \sigma_2)}{(\sigma_3 - \sigma_1)(\sigma_3 - \sigma_2)}
\end{aligned}
\tag{1.30}
$$

Without loss in generality, the principal stresses can be arranged in the order of $\sigma_1 > \sigma_2 > \sigma_3$. For the ranked principal stresses, the largest shear is easily determined as $\tau_{max} = |\tau_3| = |(\sigma_1 - \sigma_3)/2|$, and the other two shear stresses are $\tau_1 = \pm (\sigma_2 - \sigma_3)/2$ and $\tau_2 = \pm (\sigma_2 - \sigma_3)/2$, respectively.

1.6   Spherical, Deviatoric, Von Mises, and Octahedral Stresses

It is often convenient to decompose the stress into two parts, i.e., the spherical and deviatoric stress tensors. The spherical stress is defined by

$$
\sigma_0 \delta_{ij} = \begin{bmatrix} \sigma_0 & 0 & 0 \\ 0 & \sigma_0 & 0 \\ 0 & 0 & \sigma_0 \end{bmatrix}. \tag{1.31}
$$

while the deviatoric stress becomes

$$s_{ij} = \sigma_{ij} - \sigma_0 \delta_{ij}, \tag{1.32}$$

where δ_{ij} is the Kronecker δ symbol,

$$
\delta_{ij} = \begin{cases} 1, i = j \\ 0, i \neq j \end{cases}.
$$

Another special stress is related to the distortional strain energy failure criterion of plastic yielding, which is called effective or von Mises stress and given by the following expression [5]

$$\sigma_e = \sigma_{\text{von Mises}} = \sqrt{\frac{3}{2}\sigma_{ij}\sigma_{ij}}$$

$$= \frac{1}{\sqrt{2}}[(\sigma_x - \sigma_y)^2 + (\sigma_y - \sigma_z)^2 + (\sigma_z - \sigma_x)^2 + 6(\tau_{xy}^2 + \tau_{yz}^2 + \tau_{zx}^2)]^{1/2} \quad (1.33)$$

$$= \frac{1}{\sqrt{2}}[(\sigma_1 - \sigma_2)^2 + (\sigma_2 - \sigma_3)^2 + (\sigma_3 - \sigma_1)^2]^{1/2}$$

Consider a special plane whose normal makes equal angles with the three principal axes, and this plane is commonly referred to as the octahedral plane. The determinations of the normal and shear stresses are rather easy if we use the principal axes of stress. Since the unit normal vector to the octahedral plane makes equal angles with the principal axes, its components are given by $n_i = \pm(1, 1, 1)/3^{0.5}$. By using Fig. 1.11 and Eq. (1.28), it gives

$$N = \sigma_{\text{oct}} = (\sigma_1 + \sigma_2 + \sigma_3)/3 = \sigma_{ii}/3 = I_1/3$$

$$S = \tau_{\text{oct}} = \frac{1}{3}[(\sigma_1 - \sigma_2)^2 + (\sigma_2 - \sigma_3)^2 + (\sigma_3 - \sigma_1)^2]^{1/2}$$
$$= \frac{1}{3}(2I_1^2 - 6I_2)^{1/2} \quad (1.34)$$

The octahedral shear stress τ_{oct} is directly related to the distortional strain energy as well, which is often used as failure criterion of plastic yielding for ductile materials.

The relationship between von Mises stress and the octahedral shear stress is $\sigma_e = 3\tau_{\text{oct}}/2^{0.5}$.

The same treatment is valid for strain tensors as well.

References

1. Zheng XL (2000) Mechanical properties of materials, 2nd edn. Press of Northwestern Polytechnical University, Xi'am
2. Bridgman PW (1964) Studies in large plastic flow and fracture. Harvard University Press, Cambridge, Massachusetts
3. Paul SK, Roy S, Sivaprasad S, Tarafder S (2018) A simplified procedure to determine post-necking true stress-strain curve from uniaxial tensile test of round metallic specimen using DIC. J Mater Eng Perform 27(9):4893–4899
4. Hashiguchi K (2013) Elastoplasticity theory, 2nd edn. Springer, Heidelberg
5. Sadd MH (2014) Elasticity: theory, applications, and numerics, 3rd edn. Elsevier Academic Press, USA

Chapter 2
Physical Relationship in Elastoplastic Mechanics

Abstract The physical relationships in elastic mechanics, i.e., generalized Hooke's law, incremental equation and plastic yielding criteria in elastoplasticity are presented first; then, ductile damage model corresponding to the dissipation of ductility of metal is briefly described in this chapter.

2.1 Generalized Hooke's Law

2.1.1 Introduction

In Chap. 1, the major types of deformation that occur in engineering materials in room temperature is described; it includes elastic and plastic deformations. The *elastic* deformation is associated with the stretching only, instead of breaking of chemical bonds. In contrast, the inelastic deformation involves processes where atoms change their relative positions through slip of crystal planes or sliding of chain molecules. If the inelastic deformation is independent of time, it is classed as *plastic* deformation.

In actual design and analysis for engineering, the equations which present stress–strain behavior are called stress–strain relationships, or *constitutive equations*; they are inevitably needed. In elementary mechanics of materials, elastic behavior with a linear stress–strain relationship is assumed and used in calculating stresses and deflections for beams and shafts. While in more complex situations of geometry and loading, *theory of elasticity* is developed to analyze the problems by using the same basic assumptions.

In theory of elasticity, stress–strain relationships are considered in three-dimensional spaces in accordance with the actual component. In addition to elastic strains, the equations may also need to include plastic strains after plastic yielding. Regardless of the method used, analysis to determine stresses and deflections always requires appropriate stress–strain relationships for the particular material involved.

© Springer Nature Singapore Pte Ltd. 2019
M. Zheng et al., *Elastoplastic Behavior of Highly Ductile Materials*,
https://doi.org/10.1007/978-981-15-0906-3_2

The discussion of elastic deformation in three dimensions can be seen as an extension of one dimension, starting with *isotropic* behavior, where the elastic properties are the same in all directions.

2.1.2   Elastic Deformation and Elastic Constants

From the above discussions, elastic deformation is associated with stretching the chemical bonds between the atoms in a solid. Consequently, the value of the elastic modulus, E, is quite high for materials.

A *homogeneous* and *isotropic* material is the basic requirement, which means that its properties are the same at all points within the solid and in all directions. This is a simplifying assumption that is approximately true for many metals, ceramics, and amorphous materials, as well as some polymers. Although in microscopic size material composes of tiny, randomly oriented crystal grains, it is *homogeneous* and *isotropic* in macroscopic size scales.

For a *linear elastic* material, the stress is linearly related to these strains, and the correlation between stress and strain is instant and unique. In the linear elastic case, two elastic constants are needed to characterize the deformations of the material. One is the common *elastic modulus*, $E = \sigma_x/\varepsilon_x$, which is the slope of the σ_x versus ε_x line OA in Fig. 1.6. The second constant is *Poisson's ratio*, which reflects the transverse strain ε_y with respect to the longitudinal strain applied ε_x as longitudinal stress σ_x is applied,

$$v = -\frac{\varepsilon_y}{\varepsilon_x}. \tag{2.1}$$

The negative symbol in Eq. (2.1) is needed to assure a positive v since ε_y is of opposite sign to ε_x in uniaxial tension. Substituting ε_x from Eq. (2.1) into $E = \sigma_x/\varepsilon_x$ yields

$$\varepsilon_y = -v\sigma_x/E. \tag{2.2}$$

Values of the elastic modulus vary widely from material to material. Poisson's ratio is often around 0.3 and does not vary outside the range 0 to 0.5, except under very unusual circumstances. While negative values of v imply lateral expansion during axial tension, which is unlikely, as will be seen subsequently, $v = 0.5$ implies volume incompressible, and values larger than 0.5 imply a decrease in volume for tensile loading, which is also unlikely.

2.1.3 Description of Generalized Hooke's Law

Consider the general state of stress at a point. A complete description consists of normal stresses in three directions, σ_x, σ_y, and σ_z, and shear stresses on three planes, τ_{xy}, τ_{yz}, and τ_{zx}. Considering normal stresses first and assuming that small-strain case, the strains caused by each component of stress can simply be superimposed together. A stress in the x-direction causes a strain in the x-direction of σ_x/E. From Eq. (2.2), the stress σ_x could cause a strain in the y-direction of $-v\sigma_x/E$, and the same strain in the z-direction. Similarly, normal stresses in the y- and z-directions each cause strains in all three directions.

Sum up all these strains induced by normal stresses to obtain the total strain in each direction which derivates the following equations,

$$
\begin{aligned}
\varepsilon_x &= [\sigma_x - v(\sigma_y + \sigma_z)]/E \quad \text{(a)} \\
\varepsilon_y &= [\sigma_y - v(\sigma_z + \sigma_x)]/E \quad \text{(b)} \\
\varepsilon_z &= [\sigma_z - v(\sigma_x + \sigma_y)]/E \quad \text{(c)}
\end{aligned}
\tag{2.3}
$$

The shear strains that occur on the orthogonal planes are each correlated to the corresponding shear stress by the *shear modulus*, G,

$$
\gamma_{xy} = \tau_{xy}/G, \quad \gamma_{yz} = \tau_{yz}/G, \gamma_{zx} = \tau_{zx}/G. \tag{2.4}
$$

Shear strain on a given plane is only affected by the shear stresses on this plane. Equations (2.2) and (2.3) together are often called the *generalized Hooke's law*.

For an isotropic material, only two independent elastic constants are needed. E, G, and v can be considered redundant. There exists following relation among E, G, and v,

$$
G = E/[2(1 + v)]. \tag{2.5}
$$

The volumetric strain is seen to be simply the sum of the normal strains:

$$
\varepsilon_v = dV/V = \varepsilon_x + \varepsilon_y + \varepsilon_z \tag{2.6}
$$

For an isotropic material, the volumetric strain can be expressed in terms of stresses by substituting the generalized Hooke's law, specified by Eq. (2.3), into Eq. (2.6). Thus, it obtains

$$
\varepsilon_v = (1 - 2v) \cdot (\sigma_x + \sigma_y + \sigma_z)/E. \tag{2.7}
$$

The *hydrostatic stress* is defined as the average normal stress,

$$
\sigma_h = (\sigma_x + \sigma_y + \sigma_z)/3 \tag{2.8}
$$

Thus, substituting Eq. (2.8) into Eq. (2.7), it yields

$$\varepsilon_v = 3(1 - 2v)\,\sigma_h / E. \tag{2.9}$$

Equation (2.9) indicates that the volumetric strain is proportional to the hydrostatic stress. The proportional constant is called the *bulk modulus*, given by

$$K = \sigma_h / \varepsilon_v = E / [3(1 - 2v)]. \tag{2.10}$$

Additionally, there are more common constants, such as Lame's elasticity constants $\mu = G = E / [2(1 + v)]$ and $\lambda = vE / [(1 + v)(1 - 2v)] = 2vG / (1 - 2v)$.

In the following sections, it will show that ε_v and σ_h are classed as *invariant* quantities. It means that their values are always the same, which are independent of the choice of coordinate system.

Besides, if $v \to 0.5$, it causes the change in volume to be zero, $\varepsilon_v \to 0$, and $K \to \infty$; it means that the material approaches incompressibility, regardless of any kind of stresses.

2.2　Yielding Conditions in Plastic Mechanics

2.2.1　General Concept of Yielding Conditions

Engineering components may be subjected to complex loadings, such as tension, compression, bending, torsion, pressure, or combinations of these. Therefore, stresses often occur in more than one direction at a given point in the material. Such stresses can act together to cause a comprehensive strains and deformations within the material. In severe condition, the material can undergo plastic yielding or fracture. The prediction of the safe limits for use of a material under combined stresses condition needs the use of a *failure criterion*.

As the plastic yielding, there is *yielding criterion* to characterize such a phenomenon.

In Chap. 1, it has been described that the tensile specimen will enter plastic yielding status if the tensile stress reaches to the yielding strength of the specimen σ_s for an elastic–plastic one. Under this condition, the loading is uniaxial one, and it is easy to judge the plastic yielding of the specimen. However, as to a given point in the material that bears complex stresses it is not easy to judge its plastic yielding status.

The fundamental features of materials beyond the elastic deformation range are as follows [1]:

1. The inelastic deformation is irreversible or permanent.
2. The magnitude of plastic strain is larger than that of elastic strain.

3. Many experimental results indicate that the plastic yielding strength and plastic flow of conventional ductile material are unaffected by hydrostatic stress (average of normal stresses). Stress deviator or reduced stress causes initiation of plastic yielding and subsequent plastic flow. Volume of the material remains constant during plastic flow (dilatation strain is 0). Some material is suffered to permanent volume change provided it is subjected to pressure of extremely high magnitude. Under usual condition, the dilatation strain is neglected.

4. In plastic deformation range generally, the stresses and strains do not only depend on the current state of stress, but also on the entire history of loading and deformation, which means a complete path-dependent process, and there is no "one-to–one" correlation between stress and strain in plastically deformed element.

5. Only in proportional loading, the "deformation theory of plasticity" proposed by Hencky can be used.

Many criteria have been proposed to describe the initial plastic yielding condition of ductile metals. While only two plastic yielding conditions are widely accepted, which are those developed by Tresca H. and von Mises R., both conditions are independent of hydrostatic pressure.

In 1864, French Engineer Tresca H. proposed the plastic yielding condition of maximum shear stress based on a series of extrusion results, which has been accepted due to its agreement to a lot of experimental results. In 1913, German Mechanics Scientist von Mises R. proposed an equivalent stress-based yielding criterion for material. In 1924, another German Mechanics Scientist H. Hencky gave a proper interpretation to von Mises criterion. In Hencky's interpretation to von Mises criterion, he considered that if the elastically distortional energy density reaches to a critical value, the material element yields plastically.

2.2.2 Two Commonly Used Plastic Yielding Conditions

Tresca H. assumed that plastic yielding depends only on the largest shear stress in the body. With the convention, let us range an order $\sigma_1 \geq \sigma_2 \geq \sigma_3$, and thus, Tresca criterion can be expressed as $\sigma_1 - \sigma_3 = C$. The parameter C is a material constant, which can be found by experiment, such as tensile test. In a tensile test, $\sigma_3 = 0$, and thus, $\sigma_1 = \sigma_s$, i.e., the yielding strength at plastic yielding, and therefore, $C = \sigma_s$. Furthermore, Tresca criterion can be expressed as

$$\sigma_1 - \sigma_3 = \sigma_s. \tag{2.11}$$

Under the condition of pure shear, plastic yielding occurs when the largest shear stress reaches to the critical value, $\sigma_1 = k$ and $\sigma_3 = -k$, where k is the yielding strength in shear. Thus, it obtains

$$\sigma_1 - \sigma_3 = 2k. \tag{2.12}$$

Equation (2.12) is the expression of *Tresca criterion*. Equations (2.11) and (2.12) imply $k = \sigma_s/2$. Tresca criterion has been accepted due its agreement to a lot of experimental results. However, when using this criterion, it is necessary to know the magnitude and order of the principal stress so as to find the maximum shear stress τ_{max}. If the order of the principal stress can be known, it is convenient to use the Tresca criterion. From the form of its mathematical expression, it is a simple linear formula, which is very convenient to be used to solve the problem.

von Mises R. criterion postulated that plastic yielding will occur when the value of the root of mean value of the square of shear stress reaches to a critical value, which is expressed mathematically as,

$$\{[(\sigma_2 - \sigma_3)^2 + (\sigma_3 - \sigma_1)^2 + (\sigma_1 - \sigma_2)^2]/3\}^{1/2} = C_1 \tag{2.13}$$

or equivalently written as

$$(\sigma_2 - \sigma_3)^2 + (\sigma_3 - \sigma_1)^2 + (\sigma_1 - \sigma_2)^2 = C_2. \tag{2.13'}$$

Similarly, the parameter C_2 can be determined by a uniaxial tension test of a tensile specimen.

In a tension test of a tensile specimen, $\sigma_1 = \sigma_s$, $\sigma_2 = \sigma_3 = 0$ at plastic yielding; thus, von Mises criterion can be further expressed as

$$(\sigma_2 - \sigma_3)^2 + (\sigma_3 - \sigma_1)^2 + (\sigma_1 - \sigma_2)^2 = 2\sigma_s^2 = 6k^2. \tag{2.14}$$

It can be seen from Eq. (2.14) that the Mises condition satisfies the principle of stress exchange; i.e., it is no need to know the magnitude and order of the principal stress; this is an obvious advantage. In addition, it is not affected by the hydrostatic pressure; this character is the same as that of the Tresca condition. Meanwhile, Mises criterion can also give more accurate results as compared to the experimental data for ductile metallic materials.

If the left side of Eq. (2.14) is smaller than the right side, the material is considered to be in an elastic state or a rigid state. If the left side of Eq. (2.14) equals to the right side, the material is considered to enter a plastic state.

If the three coordinate axes of the σ_1, σ_2, and σ_3 are projected onto the isosceles plane of this coordinate system, the coordinates of three axes that are 120° from each other can be obtained. The results of Eq. (2.14) are shown in a coordinate of 120° to each other, and their geometry is a regular hexagon (Fig. 2.1); the results from Tresca criterion are shown as a comparison as well.

H. Hencky once studied the physical nature of von Mises criterion and gave a proper interpretation. According to the viewpoint of Hencky to von Mises criterion, the material element yields plastically if the elastically distortional strain energy density reaches to a critical value [2].

Fig. 2.1 Yield condition in
isosceles plane

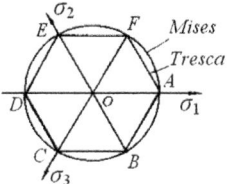

Let W, W_s, and W_v represent the total strain energy density, the distortional strain
energy density, and the dilatational strain energy density, respectively, and then

$$W = (\sigma_1\varepsilon_1 + \sigma_2\varepsilon_2 + \sigma_3\varepsilon_3)/2 = [\sigma_1^2 + \sigma_2^2 + \sigma_3^2 - 2v(\sigma_1\sigma_2 + \sigma_2\sigma_3 + \sigma_3\sigma_1)]/(2E) \tag{2.15}$$

$$W_v = (\sigma_1 + \sigma_2 + \sigma_3)(\varepsilon_1 + \varepsilon_2 + \varepsilon_3)/6 = (1 - 2v)(\sigma_1 + \sigma_2 + \sigma_3)^2/(6E), \tag{2.16}$$

$$W_s = W - W_v = [(\sigma_1 - \sigma_2)^2 + (\sigma_3 - \sigma_1)^2 + (\sigma_2 - \sigma_3)^2]/(12G), \tag{2.17}$$

in which $G = E/[2(1 + v)]$ is the shear elastic modulus.

At the same time, the total strain energy density in the element is

$$W = \frac{(1 - 2v)I_1^2}{6E} + \frac{(1 + v)J_2}{2E} = \frac{I_1^2}{18K} + \frac{J_2}{2G}, \tag{2.15'}$$

Here, I_1 and J_2 are the first stress invariant and second stress deviator invariant,
respectively.

Obviously, the expression of distortional strain energy density Eq. (2.17)
includes the left side of Eq. (2.14) (also called the octahedral shearing stress,
$\tau_{oct} = [(\sigma_1 - \sigma_2)^2 + (\sigma_2 - \sigma_3)^2 + (\sigma_3 - \sigma_1)^2]^{1/2}/3)$.

2.2.3   *Experimental Verification of Plastic Yielding Conditions*

Taylor and Quinney conducted classical experiments (combination test of pull and
twist) to verify the plastic yielding criteria. The results showed that the experimental
points fell between the two ellipses in Fig. 2.2, but were more inclined toward the
Mises ellipse. For the considered stress state, it is easy to derive the associated Lode
variable as $\mu = \sigma/(\sigma^2 + 4\tau^2)^{1/2}$.

More testing results support the plastic yielding criterion of von Mises with
sufficient evidence [2].

Besides, Zheng M. et al. once studied the effect of voids on yielding strength of
materials [3] and proposed an assessment for yielding strength of the randomly

Fig. 2.2 Comparison of yield
criteria with test results

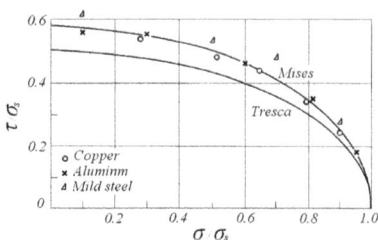

distributed void type of damaged metal based on von Mises criterion and its distortional strain energy density interpretation, which can be written as [3],

$$\sigma_s = \sigma_{s0}[(1 - \phi)(1 - 2.625\phi)/(1 + 6.25\phi\sigma_h^2/\sigma_s^2)]^{1/2}, \qquad (2.18)$$

where σ_h is the hydrostatic stress and σ_h/σ_s is called stress triaxiality.

For uniaxial tensile loading, the stress triaxiality $\sigma_h/\sigma_s = 1/3$; Eq. (1.18) becomes

$$\sigma_s = \sigma_{s0}[(1 - \phi)(1 - 2.625\phi)/(1 + 6.25\phi/9)]^{1/2}, \qquad (2.19)$$

in which ϕ is the volume fraction of voids. Equation (2.19) was verified by many available test data [3].

2.2.4 Comparison of Two Plastic Yielding Conditions

The difference between the two plastic yielding conditions is related to the method of determining the constant. If the material constant is determined by yielding limit of uniaxial tension, the more difference of two yielding conditions appears in the pure shear stress state, and the maximum shear stress determined by the Mises condition is 15.5% larger than the maximum shear stress determined by the Tresca condition; If the constant is determined by the pure shear yielding limit, the two yielding conditions differ the most in the uniaxial tension, and the maximum tensile stress determined by the Mises condition is 13.4% smaller than the maximum

Table 2.1 Comparison of features of the two plastic yielding conditions

	Difference	Similarities
von Mises	(1) Not affected by intermediate stress (2) The yield condition is nonlinear (3) No need to know the order of stress	(1) Not affected by hydrostatic pressure (2) Stress can be interchanged
Tresca	(1) Affected by intermediate stress (2) The yield function is linear (3) Need to know the order of stress	

tensile stress determined by the Tresca condition. It can be seen that the calculation results of the two yielding conditions are not much different. The similarities and differences between the two yielding conditions are listed in Table 2.1.

2.3 Relationship Between Stress and Strain in Plastic Mechanics

2.3.1 Incremental Theory in Plastic Mechanics

If a material element is unloaded from a certain plastic state, it recovers along the elastic line downwards. The elastic behavior of the material is characterized by only two independent elastic constants which retain their initial values as anisotropy is ignored. When the element is loaded again to certain stresses, plastic yielding will occur once more as long as the stress point meets the condition of plastic yielding criterion. The elastic part of the strain is always controlled by Hooke's law, and the elastic part of the strain increment is directly related to the stress increment no matter plastic yielding occurring or not. Therefore, it is reasonable to think that the increment of plastic strain is also related to the stress increment and the current stress state rationally, i.e.,

$$d\varepsilon_{ij}^p = s_{ij} \cdot d\lambda \tag{2.20}$$

in which s_{ij} is the deviatoric stress component.

An alternate form of Eq. (2.20) is

$$\frac{d\varepsilon_x^p}{s_x} = \frac{d\varepsilon_y^p}{s_y} = \frac{d\varepsilon_z^p}{s_z} = \frac{d\gamma_{xy}^p}{\tau_{xy}} = \frac{d\gamma_{yz}^p}{\tau_{yz}} = \frac{d\gamma_{zx}^p}{\tau_{zx}} = d\lambda \tag{2.21}$$

Equations (2.20) and (2.21) were the relationship to correlate the plastic strain increments to the deviatoric stress components. Lévy and von Mises independently proposed the similar response relationship for the total strain increments and the deviatoric stress component. Equation (2.21) is called Lévy-von Mises constitutive relationship.

The factor $d\lambda$ can be obtained by the loading process [3],

$$d\lambda = \frac{3d\varepsilon_{eq}^p}{2\sigma_{eq}}, \tag{2.22}$$

In Eq. (2.22), σ_{eq} and $d\varepsilon_{eq}^p$ express the equivalent flow stress and equivalent plastic strain increment, respectively. For ideal plastic material, $\sigma_{eq} = \sigma_s$.

Thus, the complete equation (Prandtl–Reuss) for an elastic–plastic material may be expressed as

$$d\varepsilon_{ij} = [(1+v)d\sigma_{ij} - v\delta_{ij} \cdot d\sigma_{kk}]/E + (\sigma_{ij} - \delta_{ij} \cdot \sigma_{kk}/3)d\lambda \qquad (2.23)$$

Equation (2.23) consists of three equations of each of the two types

$$d\varepsilon_x = [(1+v)d\sigma_x - v(d\sigma_y + d\sigma_z)]/E + (2/3) \cdot (\sigma_x - (\sigma_y + \sigma_z)/2) \cdot d\lambda, \quad (2.24)$$

$$d\gamma_{xy} = d\tau_{xy}/(2G) + \tau_{xy} \cdot d\lambda. \qquad (2.25)$$

Presently, there exist a lot of evidences to show that the von Mises plastic yielding criterion and the Prandtl–Reuss flow rule are suitable to characterize plastic behavior of most realistic metals, regardless of the anisotropy and Bauschinger effect [1].

2.3.2   Deformation Theory in Plastic Mechanics

Above discussion indicates that incremental theory of Prandtl–Reuss stress–strain relation could satisfactorily characterize plastic behavior of most realistic metals, but its incremental form makes the mathematical treatment complex. According to Hencky's simplifications, a one-to-one correspondence between the stress and the strain was developed. The components of the total plastic strain are taken to be proportional to the corresponding deviatoric stresses [1]. Under condition of small strains, the plastic stress–strain relation proposed by Hencky can be written as

$$\varepsilon_{ij}^p = \lambda s_{ij} \qquad (2.26)$$

where λ is positive during loading and zero during unloading. As to a work hardening material, λ depends on the equivalent stress σ_{eq}, which can be considered as a function of an equivalent total plastic strain ε_{eq}^p defined as $\varepsilon_{eq}^p = (2\varepsilon_{ij}^p \cdot \varepsilon_{ij}^p/3)^{1/2}$. In light of von Mises plastic yielding criterion, it gives $\lambda = 3\varepsilon_{eq}^p/2\sigma_{eq}$. Thus the stress–strain relation, Eq. (2.26) can be rewritten as

$$\varepsilon_{ij}^p = \frac{3\varepsilon_{eq}^p}{2\sigma_{eq}} s_{ij} = \frac{3}{2}\left(\frac{1}{S} - \frac{1}{E}\right)s_{ij}, \qquad (2.27)$$

where S is the secant modulus of the uniaxial stress–strain curve at $\sigma = \sigma_{eq}$. For an incompressible material, $\varepsilon_{ij}^e = 3s_{ij}/2E$ by Hooke's law; thus, Eq. (2.27) provides $\varepsilon_{ij} = 3s_{ij}/2S$. The incremental form of Eq. (2.27) is

$$d\varepsilon_{ij}^p = \frac{3}{2\sigma_{eq}} \left\{ \left(d\varepsilon_{eq}^p - \frac{\varepsilon_{eq}^p d\sigma_{eq}}{\sigma_e q} \right) s_{ij} + \varepsilon_{eq}^p ds_{ij} \right\}. \tag{2.28}$$

Hencky's equation is equivalent to the Prandtl–Reuss equation only if the ratios of the deviatoric stress components hold constant. Furthermore, Hencky's theory could be extended to large strains by using a suitable definition of the strain tensor ε_{ij}; the most common definition is $\varepsilon_{ij} = d\varepsilon_{ij}$, in which the integral is taken along the path of the particle. When the principal axes of strain increment remain fixed in the element, it could result in the logarithmic strains in the principal directions.

Hencky's theory is inappropriate to characterize the complete plastic behavior of metals, such as in consideration of unloading of the element after a certain amount of plastic deformation. If the element is reloaded to another stress state on the current yielding locus, the ratios of the plastic strain components are entirely different, and then, it is problem to deal with.

Hencky's theory is theoretically unacceptable in principle; since it differs from the plastic flow of realistic metals, it is only valid for proportional loading; in addition, there is no association between Hencky's stress–strain relation and a plastic potential with continuously turning normal [1, 2].

2.4 Ductile Damage and Fracture

2.4.1 General Description of Ductile Damage and Fracture Corresponding to Plastic Deformation

From the stress–strain curve in Fig. 1.1, it can be seen that for some materials, their plastic strain range covers a much wider part of the strain axis than their elastic range. The capacity of a material to be able to undergo such large plastic deformation is termed its *ductility*. Material with high ductility is called *ductile* material, while member with low ductility (or even no visible ductility) is termed *brittle* material. The quantitative characterization of material ductility is simply measured by its *percentage elongation* or *percentage reduction in area* with tension test [4].

It is well known that a ductile tension specimen will reach to rupture as tensile process progressing continuously. The ductile damage process accompanying the plastic deformation within the specimen is progressing during the plastic deformation. It is realized that the whole ductile damage and fracture process can be considered as three stages, i.e., micro-void formation, growth, and micro-void coalesce and microscopic crack formation leading to macroscopic failure finally [5–9]. With the development of plastic deformation, micro-void initiates at material defects (mostly inclusions) first. For some materials, micro-voids may preexist in the material in its primary status. During plastic deformation, micro-void initiates at the most weakest and favorable point of material, and these voids then grow in their particular manner related to the local stress triaxiality and increment of deformation.

The voids tend to coalesce to form microscopic crack and thus lead to macroscopic failure finally. In fact, the whole process of micro-void formation, growth, and micro-void coalesce and microscopic crack formation depend upon the actual deformation process or path that the local material element undergoes [5–9], which therefore can be reasonably characterized by the local stress triaxiality and increment of equivalent strain [5–9].

2.4.2   Ductile Damage Model Corresponding to the Dissipation of Ductility of Metal

Zheng et al. proposed a ductile damage model which corresponds to the dissipation of ductility of metal [6–9]. It considers plastic deformation dissipating the ductility of metal due to the plastic deformation, which is an irreversible process [6–9]. The formation of macrocrack results from the exhaustion of metal ductility due to large plastic deformation [6–9]. Therefore, the dissipation of ductility of metal is rationally taken as the most suitable internal damage variable to directly characterize the deterioration of the property and microstructure within deformed metal. The evolution of the damage variable D is as follows [6–9],

$$dD = A^{-1} \cdot \exp(1.5\sigma_m/\sigma_{ys}) \cdot d\varepsilon_{eq} > 0, \tag{2.29}$$

where A is a material constant characterizing the ductility of metal, σ_m is the hydrostatic stress, σ_{ys} is the instant flow stress or yielding strength, and σ_m/σ_{ys} is called stress triaxiality; $d\varepsilon_{eq}$ is the increment of equivalent strain during plastic deformation.

The material constant A can be estimated from test data of the standard cylinder specimen in tension test [6–9],

$$A = 1.65\varepsilon_u + 1.32\{1 - 0.5[\ln(1 - \psi_c) + \varepsilon_u]\}^{2.5} - 1.32, \tag{2.30}$$

where ε_u is the maximum uniform plastic strain at the center of the tensile specimen and ψ_c is the critical value of the reduction in cross-sectional area at the macrofracture of the specimen.

For tension test of a standard cylinder specimen, the stress triaxiality at the center of the specimen is the largest during the whole deformation process; therefore, damage nucleates and grows there first, which can be observed by microscope with cutting the deformed specimen at each deformation stage. However, even a crack visible to the naked eyes appears at the center of the deformed bar, the specimen can still carry load till the rupture of the specimen; i.e., the macrofracture of the standard tension test includes processes of both the onset of macrocrack initiation (visible to naked eyes) and the propagation of macrocrack till the complete rupture of the total cross section of the test bar. So, Eq. (2.30) gives the total estimation of the ductility containing both the onset of the macrocrack initiation and the propagation of macrocrack till the complete rupture of the whole specimen. Therefore, if one only prefers to determine the onset of a crack visible to naked eyes at any location of a specimen as that of the metal forming limit analysis, a modified factor, 1/3.5, was introduced to meet the needs empirically [6–9]; thus, it obtains

$$A' = 0.471\varepsilon_u + 0.377\{1 - 0.5[\ln(1 - \psi_c) + \varepsilon_u]\}^{2.5} - 0.377. \qquad (2.30')$$

The damage evolutions in upsetting process, notched bar, metal overturning, metal forming, and thin-walled torsion tests, etc., have been studied, and the results are promising [6–9].

2.4.3 Prediction of Ductile Crack Initiation Near Notch Tip Under Mode I Loading by the Ductile Damage Theory

Budden and Jones studied the distributions of stress near a crack tip that deformed in plane strain under mixed modes I and II loads by using slip-line field theory [10–12]. Their results are contrasted with both known mode I large-deformation solutions and the previous mixed-mode solutions for a sharp crack. The fracture due to the cracking mechanism addressed the micro-void growth and linkage under small-scale yielding conditions. The stress status was analyzed by taking the blunting of the crack tip into account. A damage function was defined in the intensely straining region near the blunted notch tip. The size of this damaged region extends until a critical level of damage being reached to induce a crack [10–12]. Green and Knott investigated the influences of work hardening rate and inclusion on the crack initiation and propagation in ductile low strength steels [13]. The slip-line field solution for the crack problem was obtained as well. Based on the experimental results, a model was proposed, which stated that when the strain or deformation of ligament between the crack tip and the nearest inclusion reaches to a certain critical values, a void nucleates, grows, and finally coalesces with the crack tip leading to the growth of crack [13]. Aoki et al. carried out elastic-plastic fracture tests of A5083-0 aluminum alloy under mixed mode by using compact-tension-shear method with changing the angle between the loading axis and the crack surface. Crack-tip blunting and stable crack growth were investigated by fractographic method. These experimental results were explained on the basis of the stress triaxiality near the surface and mid-thickness of the specimen [14].

Tohogo et al. studied the crack initiation from a notch and a crack of the ductile structural steel SM 41 A under mode I loading [15] by using three-point bending tests. For the notched specimens, the ductile crack initiation was predicted by the critical void volume fraction. While for the cracked specimen, a large strain field–process zone around the crack tip was introduced with void volume fraction. Finally, it suggested that the correlation between the size of large strain field and the spatial distribution of inclusions needs to be considered for the prediction of ductile crack initiation.

These results indicated that fracture behavior of metals under mixed-mode loading is intensively affected by the constraint of stress triaxiality at crack tip [15].

Zheng et al. used the ductile damage theory to analyze the ductile crack initiation near notch tip under mode I loading [16]. By consider the blunting of a crack from its initial radius of tip R_o to a new radius R as loaded, the stress triaxiality was obtained from slip-line field theory, and the damage increment at any point around the crack tip was derived. Furthermore, the onset of visible crack initiation at the crack tip was predicted with the critical blunted radius R_c based on the ductile damage theory. The comparison of the prediction with Tohogo's test data indicated the reasonability of ductile damage theory [16].

Zheng et al. also investigated the distribution and evolution of damage field in single-edge-cracked three-point bending and center-cracked tensile specimens of steel StE 690 by finite element method [17]. The stretched zone size (SZW) was proposed as the characteristic scale in this calculation. The results show that critical values of the specialized parameter based on the damage theory accumulates close to one constant though the critical J integral for physical crack initiation varies with relative crack length from one specimen to another significantly, which indicates that the damage theory is independence of the stress states and geometry of specimen [17]. An accompanying computation results showed that the local strain energy density increases with almost the same rate as crack extension regardless of specimen geometry changes [18]. It indicates that the strain energy density can be used to characterize the crack initiation for a pre-cracked specimen.

References

1. Kassir Mumtaz (2018) Applied elasticity and plasticity. CRC Press, New York
2. Chakrabarty J (2006) Theory of plasticity, 3rd edn. Elsevier, Butterworth-Heinemann
3. Zheng M, Luo ZJ, Zheng X (1992) The yielding behavior of materials with random voids. Chin Sci Bull 37:512–516
4. Hearn EJ (2000) Mechanics of materials I, an introduction to the mechanics of elastic and plastic deformation of solids and structural materials, 3rd edn. Butterworth-Heinemann, Musselburgh
5. Besson J (2010) Continuum models of ductile fracture: a review. Int J Damage Mech 19:3–52
6. Zheng M, Luo ZJ, Zheng X (1991) Correlation of notch bar, metal overturn and forming data involving effective plastic strain and stress triaxiality ratio. Theor Appl Fract Mech 16: 155–159
7. Zheng M, Luo ZJ, Zheng X (1992) A new damage model for ductile materials. Eng Fract Mech 41(1):103–110
8. Zheng M, Hu C, Luo ZJ, Zheng X (1994) Damage characterization of SAE 1020 and 1045 steel under torsion and compression. Theor Appl Fract Mech 21:91–99
9. Zheng M, Hu C, Luo ZJ, Zheng X (1996) A ductile damage model corresponding to the dissipation of ductility of metal. Eng Fract Mech 53(4):653–659
10. Budden PJ, Jones MR (1991) A ductile fracture model in mixed modes 1 and 2. Fatigue Fract Eng Mater Struct 14(4):469–482

11. Budden PJ (1987) The stress field near a blunting crack tip under mixed modes 1 and 2. J Mech Phys Solids 35(4):457–478
12. Budden PJ (1988) The effect of blunting on the strain field at a crack tip under mixed modes 1 and 2. J Mech Phys Solids 36(5):503–518
13. Green G, Knott JF (1976) The initiation and propagation of ductile fracture in low strength steels. J Eng Mater Technol 98(1):37–46
14. Aoki S, Kishimoto K, Yoshida T, Sakata M, Richard HA (1990) Elastic-plastic fracture behavior of an aluminum alloy under mixed mode loading. J Mech Phys Solids 38(2): 195–213
15. Tohogo K, Okamoto Y, Otsuk A (1988) The behavior of ductile crack initiation from a notch and a crack under mode 1 loading. JSME Int J 31(3):588–597
16. Zheng M, Zhou G, Zhao G (1995) Predicting ductile crack initiation near notch tip under mode I loading by the damage theory. Int J Fract 73:R9–R13
17. Zheng M, Lauschke U, Kuna M (2000) A damage mechanics based approach for fracture of metallic components. Comput Mater Sci 19(1–4):170–178
18. Zheng M, Zhou G, Zacharopoulos DA, Kuna M (2001) Crack initiation behavior in StE690 Steel characterized by strain energy density criterion. Theoret Appl Fract Mech 36(2): 141–145

Chapter 3
Solutions to the Typical Problem of Elastoplasticity

Abstract Solutions to the typical problems of elastoplasticity, including planar problems in elastic mechanics, stress function, examples of plane problems in Cartesian coordinates and polar coordinates, etc., are presented first; then, elastoplastic analysis and plastic limit analysis of thick-walled cylinders, elastoplastic bending of beams, and stress analysis of tube under uneven external load, etc., are presented.

3.1 Planar Problems in Elastic Mechanics

The planar problems in elasticity are defined as ones that can be dealt with in the xy-plane. Meanwhile, the shear strains in the z-plane are assumed to vanish; i.e., $\gamma_{xz} = \gamma_{yz} = 0$ for planar problems.

There are two types of planar problems: plane stress and plane strain. Plane stress problem occurs in a thin, sheet-like, or plate structure that is loaded at the edges by coplanar forces parallel to the xy-plane and uniformly distributed over the thickness, while plane strain problem treats a long member that is symmetrically loaded on the boundary at each transverse cross section. The formulations of the above two types need certain approximations so as to simplify the three-dimensional problems and reduce it to two-dimensional ones. Both problems are identical methods of solutions and mathematical formulation.

3.1.1 Plane Stress

As to an element in a thin member, the midplane of the element is taken as the xy-plane and the z-axis is the normal as shown in Fig. 3.1. The loads are applied at the boundary and uniformly distributed across its thickness. At the top and bottom surfaces, it follows $\sigma_z = \tau_{zx} = \tau_{zy} = 0$ due to the top and bottom faces of the member being free from stress; it is assumed that the same stresses vanish

© Springer Nature Singapore Pte Ltd. 2019
M. Zheng et al., *Elastoplastic Behavior of Highly Ductile Materials*,
https://doi.org/10.1007/978-981-15-0906-3_3

Fig. 3.1 Plane stress in a thin sheet

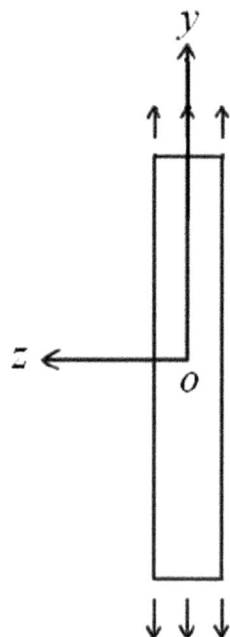

throughout the member owing to thickness of the member being ignored. The stresses within the plane of the sheet, σ_x, σ_y, and τ_{xy}, are nonvanishing and assumed to be functions of x and y.

From above assumptions, the stresses and deformations cannot satisfy all the conditions of compatibility given in Chap. 1. Therefore, in the viewpoint of pure mathematics, the plane stress solutions are not accurate. But, in the viewpoint of practical application, they are realistic provided the members are thin enough. In summary, under plane stress condition the stress distribution is determined by determining σ_x, σ_y, τ_{xy} only, and they are functions of x and y, while the stresses, σ_z, τ_{zx}, and τ_{zy}, are all vanished.

Correspondingly, the elastic strains are also independent of the variable z as well, and they could be obtained from Hooke's law as follows:

$$\varepsilon_x = (\sigma_x - v\sigma_y)/E, \varepsilon_y = (\sigma_y - v\sigma_x)/E, \gamma_{xy} = 2(1+v)\tau_{xy}/E, \varepsilon_z$$
$$= -v(\sigma_x + \sigma_y)/E, \gamma_{xz} = \gamma_{yz} = 0. \tag{3.1}$$

By adding the first two equations in Eq. (3.1), it obtains $\varepsilon_x + \varepsilon_y = (1 - v) \cdot (\sigma_x + \sigma_y)/E$, and furthermore, $\varepsilon_z = -v(\varepsilon_x + \varepsilon_y)/(1 - v)$.

If the stresses and strains are obtained, the displacement components, u, v, and w, can be gotten as well by integrating the usual strain–displacement relations, which are also functions of x and y only, $\varepsilon_x = \partial u/\partial x$, $\varepsilon_y = \partial v/\partial y, \gamma_{xy} = \partial u/\partial y + \partial v/\partial x$, and the displacement in z-direction w is obtained from the basic relation $\varepsilon_z = (\partial w/\partial z)$, which indicates that displacement w varies linearly with the depth z of the member, and usually is of no interest in thin members. Furthermore, it obtains the compatibility requirements [1],

$$\partial^2 \varepsilon_x/\partial y^2 + \partial^2 \varepsilon_y/\partial x^2 = \partial^2 \gamma_{xy}/\partial x \partial y, \partial^2 \varepsilon_z/\partial x^2 = 0, \partial^2 \varepsilon_z/\partial y^2 = 0, \partial 2\varepsilon_z/\partial x \partial y = 0. \tag{3.2}$$

The following relation is obtained, $\partial^2(\sigma_x - v\sigma_y)/\partial y^2 + \partial^2(\sigma_y - v\sigma_x)/\partial x^2 = 2(1+v)\partial^2 \tau_{xy}/\partial x \partial y$.

The equilibrium equation is [1]

$$\nabla^2(\sigma_x + \sigma_y) = 0, \tag{3.3}$$

and here, $\nabla^2 = \partial^2/\partial x^2 + \partial^2/\partial y^2$ stands for the harmonic operator in two dimensions.

When body forces are included in the equilibrium equations, then

$$\nabla^2(\sigma_x + \sigma_y) = -(1+v) \cdot (\partial F_x/\partial x + \partial F_y/\partial y). \tag{3.4}$$

Equation (3.2) is also applicable. The equilibrium equations in the presence of body forces imply that the body force F_z must vanish because there exist $\sigma_z = \tau_{xz} = \tau_{yz} = 0$.

3.1.2 Plane Strain

As to a plane strain state, it is a long prismatic member is subject to constant forces acting normal to the surface so as to keep the axial strain to be zero. If the z-axis is oriented along the length of the member, the applied forces are functions of x and y. It suffices to model a slice of unit thickness in the xy-plane as a 2D problem subject to prescribed surface loads at the boundary (Fig. 3.2). Some basic assumptions are used to reduce the three-dimensional problem into a two-dimensional (plane) one, which consist of specifying the displacement components, u, v, and w, by $u = u(x, y)$, $v = v(x, y)$, $w = 0$, and therefore, $\gamma_{xy} = \gamma_{xy}(x, y)$, $\varepsilon_z = \gamma_{xz} = \gamma_{yz} = 0$.

Furthermore, from Hooke's law, the stresses can be obtained,

$$\sigma_x = \lambda(\varepsilon_x + \varepsilon_y) + 2G\varepsilon_x, \sigma_y = \lambda(\varepsilon_x + \varepsilon_y) + 2G\varepsilon_y, \tau_{xy} = G\gamma_{xy},$$
$$\sigma_z = v(\sigma_x + \sigma_y), \tau_{xz} = \tau_{yz} = 0. \tag{3.5}$$

Fig. 3.2 Plane strain in a
long structure

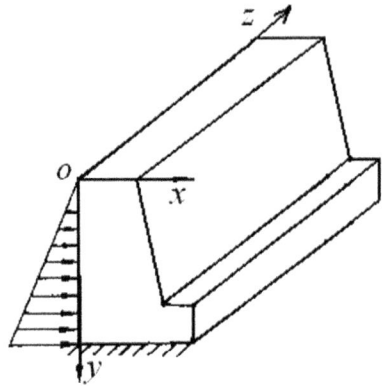

Correspondingly, the nonzero strains are given by

$$\varepsilon_x = (1+v)[(1-v)\sigma_x - v\sigma_y]/E, \varepsilon_y = (1+v)[(1-v)\sigma_y - v\sigma_x]/E, \gamma_{xy}$$
$$= 2(1+v)\tau_{xy}/E. \tag{3.6}$$

Now, the problem involved in plane strain is that the structure is subjected to axial strain at all points of the cross section, and axial stress, σ_z, in the above case is added.

Furthermore, the compatibility requirement for plane strain problems in terms of stresses is $\nabla^2(\sigma_x + \sigma_y) = 0$ again.

When body forces are present, the compatibility equation in terms of stresses becomes

$$\nabla^2(\sigma_x + \sigma_y) = -(\partial F_x/\partial x + \partial F_y/\partial y)/(1-v). \tag{3.7}$$

Compared Eq. (3.4) with Eq. (3.7), it can be summarized that the stresses in plane stress and plane strain problems are controlled by identical compatibility equations. In order to determine the stresses in plane stress and plane strain problems, one needs to solve the differential equations of equilibrium together with the compatibility equations subjected to the proper boundary conditions. Once the stresses are determined, the strains are obtained from Eqs. (3.1) and (3.6), and the displacement components are then obtained from the strain–displacement relations. Comparison of Eqs. (3.1) and (3.6) indicates that converting the strains from plane stress into plane strain, one needs to replace E and v in plane stress by $E/(1-v^2)$ and $v/(1-v)$ for plane strain, respectively. Conversely, to convert from plane strain into plane stress, one needs to substitute for the constants E and v in plane strain by $((1+2v)E)/(1+v)^2$ and $v/(1+v)$ for plane stress, respectively.

3.2 Stress Function

In the absence of body forces, plane stress and plane strain problems are governed by two equilibrium equations [1]

$$\partial\sigma_x/\partial x + \partial\tau_{xy}/\partial y = 0, \quad \partial\tau_{xy}/\partial x + \partial\sigma_y/\partial y = 0. \tag{3.8}$$

The two compatibility relations, i.e., Eqs. (3.4) and (3.7), now become

$$\nabla^2(\sigma_x + \sigma_y) = 0 \tag{3.9}$$

and the appropriate boundary conditions. In order to get the stresses, a new function of x and y, $\varphi(x, y)$, is introduced, such that the equilibrium equations are satisfied automatically. Such function was first introduced by Airy in 1863, so it is called Airy stress function. The assumptions are

$$\sigma_x = \partial^2\varphi/\partial y^2, \sigma_y = \partial^2\varphi/\partial x^2, \tau_{xy} = -\partial^2\varphi/\partial x\partial y. \tag{3.10}$$

It is readily confirmed that Eq. (3.8) is automatically satisfied. With the compatibility equation, a fourth-order partial differential equation is found that it governs stress function,

$$\partial^4\varphi/\partial x^4 + 2\partial^4\varphi/\partial x^2\partial y^2 + \partial^4\varphi/\partial y^4 = 0, \text{ or } \nabla^4\varphi = 0 \tag{3.11}$$

where ∇^4 the bi-harmonic operator stands for the abbreviation $\nabla^4 = \partial^4/\partial x^4 + 2\partial^4/\partial x^2\partial y^2 + \partial^4/\partial y^4$.

The stress function, φ, is determined by solving Eq. (3.11), which satisfies appropriate boundary conditions of the problem.

Boundary conditions: At the boundary, the stresses must be in equilibrium with the external forces present at that part. If the components of the external load per unit area are expressed by p_x and p_y, the direction cosines of the normal to the boundary surface are denoted by l and m, respectively. Equilibrium of the element yields

$$p_x = \sigma_x \cdot l + \tau_{xy} \cdot m, \quad p_y = \tau_{xy} \cdot l + \sigma_y \cdot m. \tag{3.11}$$

Body force: In the presence of body forces, similar procedures can be used. Assume that the body force is conservative with a potential of $V(x, y)$. Then, the components of the body force can be written as, $F_x = -\partial V/\partial x, F_y = -\partial V/\partial y$. Thus, the equilibrium equations now become

$$\partial(\sigma_x - V)/\partial x + \partial\tau_{xy}/\partial y = 0, \partial\tau_{xy}/\partial x + \partial(\sigma_y - V)/\partial y = 0. \tag{3.12}$$

If we introduce $\sigma_x - V = \partial^2\varphi/\partial y^2$, $\sigma_y - V = \partial^2\varphi/\partial x^2$, $\tau_{xy} = -\partial^2\varphi/\partial x\partial y$, then the compatibility equation for plane stress problems is

$$\nabla^4\varphi = -(1+v)\nabla^2 V \qquad\qquad (3.13)$$

Similarly, the equation for plane strain problems is

$$\nabla^4\varphi = -[1/(1+v)]^2 \cdot \nabla^2 V \qquad\qquad (3.13)$$

Solutions of planar elasticity problems: The planar elasticity problems solving in plane elastic bodies is to develop solutions of the bi-harmonic equation such that the stress components could satisfy the appropriate boundary conditions.

3.3   Example of Plane Problem in Cartesian Coordinates

Figure 3.3 shows an example; a long rectangular cylinder with a material density of ρ is placed in a rigid groove of the same shape. If friction is ignored, the form of the stress function is $\varphi = Ax^2y + By^3 + Cy^2 + Dx^2$. Try to give the stress components, the strain components, and the displacement values on the boundary [2].

Based on the given stress function and the Airy stress function, it obtains

$$\sigma_x = \partial^2\varphi/\partial y^2 = 6By + 2C, \sigma_y = \partial^2\varphi/\partial x^2 = 6Ay + 2D, \tau_{xy} = -\partial^2\varphi/\partial x\partial y + \rho x$$
$$= -2Ax + \rho x.$$

At the boundary $y = h$: (1) $\tau_{xy} = 0$; (2) $\sigma_y = 0$. And the boundary condition for rigid groove is $\int_{-a}^{a} \varepsilon_x \cdot dx = 0$, and it results in $\varepsilon_x = 0$ due to ε_x itself being independent of x.

Fig. 3.3 A long rectangular cylinder in a rigid groove

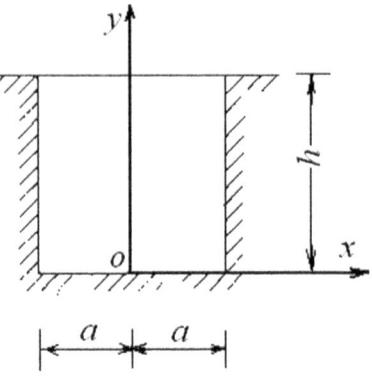

By using the boundary condition (1), it yields $-2Ax + \rho x = 0$, and thus, $A = \rho/2$. While for the boundary condition (2), it yields $2h \cdot \rho/2 + 2D = 0$, and thus, $D = -h \cdot \rho/2$. The boundary condition (3) leads to $\varepsilon_x = (1 + v)[(1 - v)\sigma_x - v\sigma_y]/E$, and thus, $\sigma_x = v\sigma_y/(1 - v)$. Furthermore, $\sigma_{y=}\rho y - \rho h$, so $\sigma_x = v\rho(y - h)/(1 - v)$. Thus, $B = v/[6(1 - v)]$, $C = -v\rho h/[2(1 - v)]$. $\varepsilon_y = (1 + v)(1 - 2v)$ $\rho(y - h)/[E(1 - v)]$. The displacements u and v can be determined by the geometric relations, $u = \int \varepsilon_x \cdot dx = 0$, $v = \int \varepsilon_y \cdot dx = \frac{(1+v)(1-2v)}{E(1-v)}\rho\left(\frac{y^2}{2} - hy\right) + K$; if $y = 0$, and $v = 0$, this results in the integral constant $K = 0$. Thus, it yields the displacement at $y = h$, $v|_{y=h} = -\frac{(1+v)(1-2v)}{2E(1-v)}\rho h^2$.

3.4 The Plane Problems in Polar Coordinates

As to practical structure and problem, the shape of structure and the force applied on it are complex, and the selection of appropriate coordinate system will make the problem solving convenient. For example, for cylinders, disks, and sector plates, as well as semi-planar bodies, polar coordinates are convenient.

In the plane problem, stress, strain, and displacement are only functions of r and θ, but independent of the axial z, the polar coordinate is a special case of the basic equation of cylindrical coordinates, and the physical equation is

$$\varepsilon_r = (\sigma_r - v\sigma_\theta)/E, \varepsilon_\theta = (\sigma_\theta - v\sigma_r)/E, \gamma_{r\theta} = 2(1 + v)\tau_{r\theta}/E. \tag{3.14}$$

For the plane strain problem, it is only necessary to replace E and E_1 in the above equations with $E_1 = E/(1 - v^2)$ and $v_1 = v/(1 - v)$, respectively.

For the axisymmetric problem, the stresses and strains are independent of θ, and thus, the problem is transformed into one-dimensional problems; such problems have a crucial meaning in the engineering process.

Since the stress is only a function of r, the bi-harmonic equation can be reduced to an ordinary differential equation; i.e.,

$$\nabla^4\varphi = [d^2/dr^2 + (1/r) \cdot d/dr] \cdot [d^2\varphi/dr^2 + (1/r) \cdot d\varphi/dr] = 0. \tag{3.15}$$

Or written as

$$\frac{1}{r} \cdot \frac{d}{dr}\left\{r\frac{d}{dr}\left[\frac{1}{r} \cdot \frac{d}{dr}\left(r \cdot \frac{d\varphi}{dr}\right)\right]\right\} = 0 \tag{3.16}$$

By integrating Eq. (3.16) directly, it yields such a stress function

$$\varphi = A \ln r + B r^2 \ln r + C r^2 + D, \tag{3.17}$$

wherein A, B, C, and D are integral constants.

Thus, stress components, strain components, and displacements can be written as,

$$\sigma_r = (1/r) \cdot \mathrm{d}\varphi/\mathrm{d}r = A/r^2 + 2C, \sigma_\theta = \mathrm{d}^2\varphi/\mathrm{d}r^2 = -A/r^2 + 2C, \tau_{r\theta}$$
$$= -\partial[(1/r) \cdot \partial\varphi/\partial\theta]/\partial r = 0 \tag{3.18}$$

$$\varepsilon_r = \frac{1}{E}[(1+v)\frac{A}{r^2} + 2(1-v)C], \quad \varepsilon_\theta = \frac{1}{E}[-(1+v)\frac{A}{r^2} + 2(1-v)C], \quad \gamma_{r\theta} = 0. \tag{3.19}$$

$$u = \frac{1}{E}[-(1+v)\frac{A}{r} + 2(1-v)Cr], v = 0. \tag{3.20}$$

Because u is independent of θ, and $v = 0$, the integral constants including B in the integral can be selected as zero.

Example 1 A wedge of unit thickness as shown in Fig. 3.4 is known, and its central angle is 2α. At the tip end of the wedge, the action of the concentrated force P is also known. Then, it needs to get expression of the strain components.

Solution From the character of the stress function on the boundary, point o can be taken as the initial point, and let $\varphi|_o = 0$, and then the moment along the *om* segment of the boundary is $\varphi = M = Pr\sin(\beta - \alpha)$. So the solution of φ is assumed as $\varphi = rf(\theta)$, substituting this formula into the harmonic equation, it yields $\frac{\mathrm{d}^4 f}{\mathrm{d}\theta^4} + 2\frac{\mathrm{d}^2 f}{\mathrm{d}\theta^2} + f = 0$. Finally, it obtains $\varphi(r, \theta) = r\theta(C\cos\theta + D\sin\theta)$, and $\sigma_r = -2(C\sin\theta - D\cos\theta)/r, \sigma_\theta = 0, \tau_{r\theta} = 0$.

Fig. 3.4 Wedge and its stress distribution

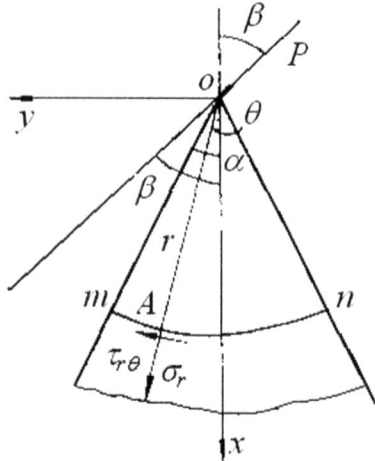

By using the boundary condition on mn [2], it yields $C = P \sin \beta /(2\alpha - \sin 2\alpha)$, $D = P \cos \beta /(2\alpha + \sin 2\alpha)$, and $\sigma_r = -\frac{2P \sin \beta \sin \theta}{r(2\alpha - \sin 2\alpha)} - \frac{2P \cos \beta \cos \theta}{r(2\alpha + \sin 2\alpha)}$, $\sigma_\theta = 0, \tau_{r\theta} = 0$.

Example 2 It is known that a rectangular cross-sectional curved rod is subjected to the bending moment M, determine the stress function and the corresponding stress component by using the mechanical function of the stress function method and its derivative (Fig. 3.5) [2].

Solution Set A as the starting point, and let $\varphi_A = 0, (\partial\varphi/\partial r)_A = 0, (\partial\varphi/\partial\theta)_A = 0$. On the boundary $(r = a), \varphi = 0, (\partial\varphi/\partial r)_{r=a} = 0$, and $(\partial\varphi/\partial\theta)_{r=a} = 0$. On the boundary $(r = b), \varphi = M, (\partial\varphi/\partial r)_{r=b} = 0, (\partial\varphi/\partial\theta)_{r=b} = 0$. Thus, φ is independent of r and θ on the boundaries, so it assumes $\varphi = f(r)$. Substituting it into the harmonic equation, it yields $\frac{d^4f}{dr^4} + \frac{2}{r}\frac{d^3f}{dr^3} - \frac{1}{r^2}\frac{d^2f}{dr^2} + \frac{1}{r^3}\frac{df}{dr} = 0$. Let $f(r) = r^n$, and substituting it into the harmonic equation, it obtains $n^2(n-2)^2 \cdot r^{n-4} = 0$, i.e., $n^2(n-2)^2 = 0$ with the solutions of $n_{1,2} = 0$ and $n_{3,4} = 2$. So it obtains $f(r) = C_1 r^2 + C_2 r^2 \ln r + C_3 \ln r + C_4$. The descriptions of stresses are $\sigma_r = 2C_1 + C_2(2\ln r + 3) - C_3/r^2, \tau_{r\theta} = 0$. By using the boundary condition at $r = a$, $\varphi = 0, \partial\varphi/\partial r = 0$, and $r = b, \varphi = M, \partial\varphi/\partial r = 0$, it yields $C_1 a^2 + C_2 a^2 \ln a + C_4 = 0$, $2C_1 + C_2(\ln a + 1) + C_3/a^2 = 0, C_1 b^2 + C_2 b^2 \ln b + C_3 \ln b + C_4 = M, 2C_1 + C_2(2\ln b + 1) + C_3/b^2 = 0$. The solutions are $\sigma_r = -\frac{M}{N}\left(b^2 \ln\frac{r}{b} + a^2 \ln\frac{a}{r} + \frac{a^2 b^2}{r^2}\ln\frac{b}{a}\right)$, $\sigma_\theta = -\frac{4M}{N}\left(b^2 - a^2 + b^2 \ln\frac{r}{b} + a^2 \ln\frac{a}{r} - \frac{a^2 b^2}{r^2}\ln\frac{b}{a}\right), \tau_{r\theta} = 0$, with $N = (b^2 - a^2)^2 - 4a^2 b^2 \left(\ln\frac{b}{a}\right)^2$.

Example 3 The stress distribution in a thin rectangular plate with a circular hole under the condition of a uniaxial tension q is a classical problem. While in the absence of circular hole, the stress distribution in the thin rectangular plate is uniform, as in Fig. 3.6a. However, as a circular hole is introduced in the plate, it disturbs the uniform distribution of stress near the hole and results in a significantly higher tensile stress near the hole, shown in Fig. 3.6b [2].

Solution By introducing the stress function, $\varphi(r, \theta) = \varphi_1(r) + \varphi_2(r, \theta)$; here, $\varphi_1(r)$ represents symmetric part of the stress distribution, and $\varphi_2(r, \theta)$ is uniaxially tensile part. Set a bigger circle with the radius b, and outside this circle, the disturbance and the stress concentration are rather small and can even be ignored. At remote position, the stress distributions approach those of the uniaxially tensile condition of a rectangular plate, i.e., $\sigma_x \to q, \sigma_y \to 0, \tau_{r\theta} \to 0$.

Fig. 3.5 Curved rod with bending moment

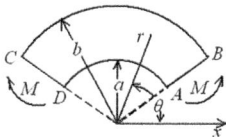

Fig. 3.6 Stress distribution
in a thin rectangular plate with
a circular hole

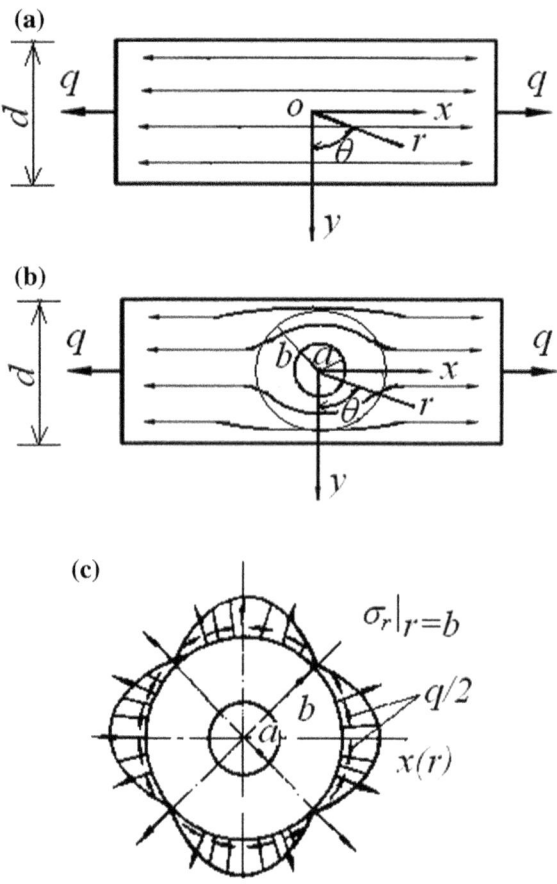

According to the symmetric characteristics of the problem, at a any radius
$r = a, \sigma_r^{(1)} = \tau_{r\theta}^{(1)} = 0$, and at $r = b, \sigma_r^{(1)} = q/2, \tau_{r\theta}^{(1)} = 0$. Therefore, the partial
solutions could be $\sigma_r^{(1)} = (1 - a^2/r^2)q/2, \sigma_\theta^{(1)} = (1 + a^2/r^2)q/2, \tau_{r\theta}^{(1)} = 0$.

As to $\varphi_2(r, \theta)$, its boundary conditions are as follows: at $r = a, \sigma_r^{(2)} = \tau_{r\theta}^{(2)} = 0$,
and at $r = b, \sigma_r^{(2)} = q\cos(2\theta)/2, \tau_{r\theta}^{(2)} = -q\sin(2\theta)/2$. Additionally, $\varphi_2(r,\theta)$ satisfies the harmonic equation, i.e., $\nabla^2\nabla^2\varphi_2(r,\theta) = 0$.

According to the symmetric characteristics of the problem and the type of the
load at remotes position being $\cos 2\theta$, it assumes $\varphi_2(r,\theta) = f(r)\cos 2\theta$; thus, it
obtains

$$r\frac{d}{dr}\left\{\frac{1}{r^2}\frac{d}{dr}\left\{r^3\frac{d}{dr}\left[\frac{1}{r^3}\frac{d}{dr}\left(r^2 f(r)\right)\right]\right\}\right\} = 0.$$

After integrating, it yields

$$\varphi_2(r, \theta) = \left(Ar^4 + Br^2 + C + D/r^2\right)\cos 2\theta.$$

The four integral constants can be determined by four boundary conditions and the finiteness at $a/b \to 0$; it results in $A = 0, B = -q/4, C = a^2q/2$, and $D = -a^4q/4$.

Thus, the partial stresses are

$$\sigma_r^{(2)} = \frac{q}{2}\left(1 - \frac{a^2}{r^2}\right)\left(1 - 3\frac{a^2}{r^2}\right)\cos 2\theta$$

$$\sigma_\theta^{(2)} = -\frac{q}{2}\left(1 + 3\frac{a^4}{r^4}\right)\cos 2\theta$$

$$\tau_{r\theta}^{(2)} = -\frac{q}{2}\left(1 - \frac{a^2}{r^2}\right)\left(1 + 3\frac{a^2}{r^2}\right)\sin 2\theta$$

Finally, the whole stress distribution of this problem is

$$\sigma_r = \frac{q}{2}\left(1 - \frac{a^2}{r^2}\right) + \frac{q}{2}\left(1 - \frac{a^2}{r^2}\right)\left(1 - 3\frac{a^2}{r^2}\right)\cos 2\theta$$

$$\sigma_\theta = \frac{q}{2}\left(1 + \frac{a^2}{r^2}\right) - \frac{q}{2}\left(1 + 3\frac{a^4}{r^4}\right)\cos 2\theta$$

$$\tau_{r\theta} = -\frac{q}{2}\left(1 - \frac{a^2}{r^2}\right)\left(1 + 3\frac{a^2}{r^2}\right)\sin 2\theta.$$

From the above solutions, at $r = a$, $\sigma_\theta = q[1 - 2\cos(2\theta)]$, thus at $\theta = \pm\pi/2$, $\sigma_\theta = 3q$, which means the *stress concentration factor at the hole edge is* 3; at $\theta = \pm\pi/2$, $\sigma_\theta = -q$, the stress at this point behaves compressive one. At $\theta = \pm\pi/2$, $\sigma_\theta = q\left(1 + \frac{a^2}{2r^2} + 3\frac{a^4}{r^4}\right)$, the stress decreases with r rapidly, if $r = a$, $\sigma_\theta = 3q$; $r = 2a$, $\sigma_\theta = 1.22q$; $r = a$, $\sigma_\theta = 1.07q$. It is obvious that the stress decays quite fast, Fig. 3.5c [2].

3.5   Elastoplastic Analysis and Plastic Limit Analysis of Thick-Walled Cylinders

Consider a hollow thick-walled cylinder under the action of uniform internal and external pressure loadings, as shown in Fig. 3.7a. It assumes that the cylinder is long, and thus, this problem can be solved with plane strain model [2].

Take the polar coordinates (r, θ), since the symmetric property of the problem, $\tau_{r\theta} = 0$; the radial and tangential stresses $\sigma_r(r)$ and $\sigma_\theta(r)$ are only functions of r while independence of θ. Similarly, the same is true for $\varepsilon_r(r)$ and $\varepsilon_\theta(r)$; correspondingly, there is only radial displacement $u(r)$; the axial displacement depends on z only, i.e., $w = w(z)$.

The equilibrium equation is

$$\frac{d\sigma_r}{dr} + \frac{\sigma_r - \sigma_\theta}{r} = 0. \tag{3.21}$$

The geometric equations are

$$\varepsilon_r = \frac{du}{dr}, \varepsilon_\theta = \frac{u}{dr}. \tag{3.22}$$

The constitutive equations for plane stress condition are

$$\varepsilon_r = \frac{1}{E}(\sigma_r - v\sigma_\theta), \varepsilon_\theta = \frac{1}{E}(\sigma_\theta - v\sigma_r). \tag{3.23}$$

As to plane strain condition, one needs only replace E and v in Eq. (3.23) by $E_1 = 1/(1 - v^2)$ and $v_1 = 1/(1 - v)$.

(1) *Elastic Case*

The nonzero stresses are given by the form, $\sigma_r = A/r^2 + B$, $\sigma_\theta = -A/r^2 + B$.

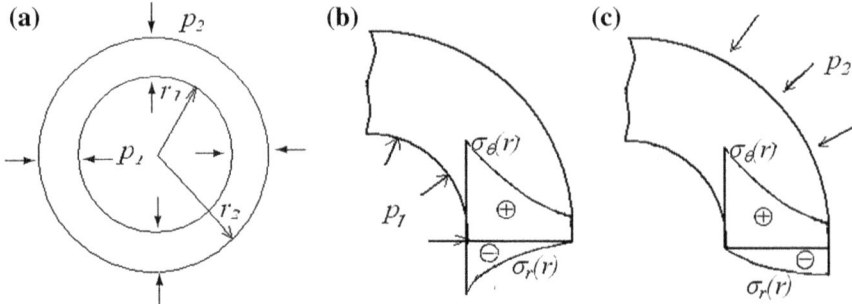

Fig. 3.7   Thick-walled cylinder

The boundary conditions are $\sigma_r(r_1) = -p_1$, $\sigma_r(r_2) = -p_2$; thus, the two unknown constants A and B can be determined, $A = r_1^2 \cdot r_2^2(p_2 - p_1)/(r_2^2 - r_1^2)$, $B = (r_1^2 \cdot p_1 - r_2^2 \cdot p_2)/(r_2^2 - r_1^2)$. Thus, the elastic stress field is

$$\sigma_r = \frac{r_1^2 r_2^2 (p_2 - p_1)}{r_2^2 - r_1^2} \frac{1}{r^2} + \frac{r_1^2 p_1 - r_2^2 p_2}{r_2^2 - r_1^2}, \quad \sigma_\theta = -\frac{r_1^2 r_2^2 (p_2 - p_1)}{r_2^2 - r_1^2} \frac{1}{r^2} + \frac{r_1^2 p_1 - r_2^2 p_2}{r_2^2 - r_1^2}.$$
$$(3.24)$$

As to the very long thick-walled cylinder, according to plane strain theory, the out-of-plane longitudinal stress is given by $\sigma_z = v(\sigma_r + \sigma_\theta) = 2v \frac{r_1^2 p_1 - r_2^2 p_2}{r_2^2 - r_1^2}$.

The radial displacement can be obtained by using Hooke's law and strain–displacement relations as

$$u_r = \frac{1+v}{E} r \left[(1 - 2v)B - \frac{A}{r^2} \right]$$
$$= \frac{1+v}{E} \left[-\frac{r_1^2 r_2^2 (p_2 - p_1)}{r_2^2 - r_1^2} \frac{1}{r} + (1 - 2v) \frac{r_1^2 p_1 - r_2^2 p_2}{r_2^2 - r_1^2} r \right]. \quad (3.25)$$

The two special cases are as follows. In the absence of outside pressure, i.e., only internal pressure case, $p_2 = 0$,

$$\sigma_r = -\frac{r_1^2 r_2^2 p_1}{r_2^2 - r_1^2} \frac{1}{r^2} + \frac{r_1^2 p_1}{r_2^2 - r_1^2}, \quad \sigma_\theta = \frac{r_1^2 r_2^2 p_1}{r_2^2 - r_1^2} \frac{1}{r^2} + \frac{r_1^2 p_1}{r_2^2 - r_1^2}. \quad (3.26)$$

The radial stress decays from $-p_1$ to zero outward rapidly from the inner radius, while the hoop stress exhibits always positive with a maximum value at the inner radius $\sigma_\theta(r_1) = \frac{(r_2^2 + r_1^2)p_1}{r_2^2 - r_1^2}$, Fig. 3.7b. In the absence of internal pressure, i.e., only outer pressure case, $p_1 = 0$,

$$\sigma_r = \frac{r_1^2 r_2^2 p_2}{r_2^2 - r_1^2} \frac{1}{r^2} - \frac{r_2^2 p_2}{r_2^2 - r_1^2}, \quad \sigma_\theta = -\frac{r_1^2 r_2^2 p_2}{r_2^2 - r_1^2} \frac{1}{r^2} - \frac{r_2^2 p_2}{r_2^2 - r_1^2}. \quad (3.27)$$

The radial stress reduces from $-p_2$ to zero rapidly inward from the outer radius, while the hoop stress exhibits always negative with a minimum value at the outer radius $\sigma_\theta(r_2) = -p_2 \frac{(r_2^2 + r_1^2)p_1}{r_2^2 - r_1^2}$, Fig. 3.7c.

(2) *Plastic Case*

As discussed in the previous section, in the absence of outside pressure, i.e., only internal pressure case, $p_2 = 0$, the stresses' distributions are

$$\sigma_r = -\frac{r_1^2 r_2^2 p_1}{r_2^2 - r_1^2}\frac{1}{r^2} + \frac{r_1^2 p_1}{r_2^2 - r_1^2}, \ \sigma_\theta = \frac{r_1^2 r_2^2 p_1}{r_2^2 - r_1^2}\frac{1}{r^2} + \frac{r_1^2 p_1}{r_2^2 - r_1^2}.$$

If the inner pressure is small, the thick wall cylinder is in elastic status, and the maximum of $|\sigma_\theta(r) - \sigma_r(r)|$ occurs at the inner radius $r = a$ with the value of $2r_2^2 p_1/(r_2^2 - r_1^2)$. Therefore, with the increase of inner pressure p_1, plastic yielding may occur first at the inner radius. The basic condition for the inner radius plastic yielding is $|\sigma_\theta(r) - \sigma_r(r)| = \sigma_s$; here, σ_s is the plastic yielding strength of the cylinder material. Thus, the elastic limit inner pressure for the initial plastic yielding occurring at the inner radius is obtained

$$p_{1e} = \sigma_s(1 - r_1^2/r_2^2)/2. \tag{3.28}$$

When $p_1 < p_{1e}$, the cylinder is in elastic status; when $p_1 > p_{1e}$, the cylinder is in plastic status near the inner radius, and the plastic zone expands outward with the increase of inner pressure. Take the interface of the elastic and plastic zones at r_p, and the whole thick-walled cylinder can be divided into two deformation zones. The inner cylinder follows the plastic yielding rule, while the elastic principle holds in the outer cylinder. The radial uniform pressure q is applied on the interface, $\sigma_r|_{r=r_p} = q$. In the plastic one, the stresses follow equilibrium equation and plastic yielding condition simultaneously, i.e.,

$$\mathrm{d}\sigma_r/\mathrm{d}r + (\sigma_r - \sigma_\theta)/r = 0, \quad \sigma_\theta - \sigma_{r=\sigma_s}. \tag{3.29}$$

Substitute the plastic yielding condition into equilibrium equation, it yields

$$\mathrm{d}\sigma_r/\mathrm{d}r - \sigma_s/r = 0, \mathrm{d}\sigma_r = \sigma_s \mathrm{d}r/r, \sigma_r = \sigma_s lnr + C. \tag{3.30}$$

The integral constant C can be determined at the inner boundary, $r = r_1$, $\sigma_r|_{r=r_p} = -p_p$; thus, it derives $C = -p_p - \sigma_s \ln a$. Furthermore, it could obtain the expression for in the plastic zone. Finally, the total stress distributions in the plastic zone can be written as,

$$\sigma_r = \sigma_s \ln(r/a) - p_p, \sigma_\theta = \sigma_s[1 + \ln(r/a)] - p_p. \tag{3.31}$$

Equation (3.31) indicates that the stress components are statically determined and dependence of the inner pressure in the plastic zone, and $\sigma_r < 0$, $\sigma_\theta > 0$.

While in the elastic zone, it is now a cylinder with the inner radius r_p and outer radius r_2 and suffered the inner pressure q. Equation (3.26) becomes

$$\sigma_r = -\frac{r_p^2 r_2^2 q}{r_2^2 - r_p^2}\frac{1}{r^2} + \frac{r_p^2 q_1}{r_2^2 - r_p^2}, \ \sigma_\theta = \frac{r_p^2 r_2^2 q}{r_2^2 - r_p^2}\frac{1}{r^2} + \frac{r_p^2 q}{r_2^2 - r_p^2}. \tag{3.32}$$

In Eq. (3.32), the values q and r_p are undetermined, which can be determined by the boundary condition at the interface of elastic and plastic zones. In the elastic zone, its inner surface $r = r_p$ starts plastic yielding; thus, $q = \sigma_s(1 - r_p^2/r_2^2)$. As to the plastic zone, its outer surface is at $r = r_p$, and $\sigma_r|_{r=rp} = -q$, substituting this result into the first formula of Eq. (3.31), it yields $q = p_p - \sigma_s\ln(r_p/r_1)$. Furthermore, from the condition of radial pressure being equal of the both sides at $r = r_p$,

$$p_p = \sigma_s ln(r_p/r_1) + \sigma_s(1 - r_p^2/r_2^2)/2. \tag{3.33}$$

Equation (3.33) indicates that the radius of the interface r_p can be determined by p_p or vice versa.

Furthermore, substituting the value of q obtained from the plastic yielding zone into the stress distributions of the elastic zone, it yields the total distributions of stresses in the elastic zone ($r_p \ll r_2$),

$$\sigma_r = -\frac{\sigma_s r_p^2}{2r_2^2}\left(\frac{r_2^2}{r^2} - 1\right), \; \sigma_\theta = \frac{\sigma_s r_p^2}{2r_2^2}\left(\frac{r_2^2}{r^2} + 1\right). \tag{3.34}$$

The plastic zone expands with the increase of the inner pressure. If the plastic zone expands to the outer radius r_2, i.e., $r_p = r_2$, the whole cylinder enters plastic yielding status; thus, a critical inner pressure is obtained; i.e.,

$$p_l = \sigma_s \ln(r_2/r_1). \tag{3.35}$$

Under condition of the whole cylinder entering plastic yielding status, $p_p = p_l$, the stress distributions are

$$\sigma_r = \sigma_s \ln(r/r_2), \sigma_\theta = \sigma_s[1 + ln(r/r_2)]. \tag{3.36}$$

In summary, under three conditions of elastic, elastic–plastic, and complete plastic status, there always exist $\sigma_r < o$, $\sigma_\theta > 0$. The maximum of absolute value of σ_r occurs at the inner surface, while maximum of absolute value of σ_θ moves outward from the inner surface with the increase of inner pressure.

In plane strain case and under plastic condition, $\varepsilon_z = 0$, it derives $\sigma_z = (\sigma_r + \sigma_\theta)/2$; if von Mises yielding criterion is employed, it yields

$$\sigma_\theta - \sigma_r = 2\sigma_s/3^{0.5} = 1.155\sigma_s. \tag{3.37}$$

On the other hand, if Tresca yielding criterion is used, it obtains

$$\sigma_\theta - \sigma_r = \sigma_s. \tag{3.38}$$

The difference of $\sigma_\theta - \sigma_r$ for the two yielding criteria is 15.5%.

(3) *Displacement*

The elastic zone is ranges $r_p < r < r_2$, and the internal pressure q, the distribution of displacement is [2],

$$u = \frac{(1+v)r_p^2 q}{E(r_2^2 - r_p^2)}\left[(1 - 2v)r + \frac{b^2}{r}\right].$$
(3.39)

The q can be considered as the elastic limit of the elastic zone, so

$$u = \frac{(1+v)\sigma_s r_p^2}{2Er_2^2 r}\left[(1 - 2v)r^2 + r_2^2\right], \quad r_p < r < r_2.$$
(3.40)

In plastic zone, considering conditions of plane strain $\varepsilon_z = 0$ and the incompressible volume $\varepsilon_r + \varepsilon_\theta = 0$, it yields

$$\frac{du}{dr} + \frac{u}{r} = 0, \text{ or } \frac{1}{r}\frac{d}{dr}(ru) = 0,$$
(3.41)

Integrating Eq. (3.41) yields

$$u = C_1/r.$$
(3.42)

The constant C_1 can be determined by the continuous condition at the interface of elastic and plastic zones, $C_1 = \frac{(1+v)\sigma_s r_p^2}{2Er_2^2}\left[(1 - 2v)r_p^2 + r_2^2\right]$, and then

$$u = \frac{(1+v)\sigma_s r_p^2}{2Er_2^2 r}\left[(1 - 2v)r_p^2 + r_2^2\right], \quad r_1 < r < r_p.$$
(3.43)

If take $v = 1/2$, i.e., incompressible volume in elastic case as well, Eqs. (3.40) and (3.43) yield the same result

$$u = \frac{3}{4}\frac{\sigma_s r_p^2}{Er}.$$
(3.44)

Equation (3.44) indicates that the radial displacement u gets its maximum at the inner surface $r = r_1$, $u = \frac{3}{4}\frac{\sigma_s r_p^2}{Er_1}$. When $r_p = r_1$, i.e., elastic limit of the inner surface, $u_e = \frac{3}{4}\frac{\sigma_s r_1}{E}$; when $r_p = r_2$, i.e., the complete plastic yielding of the whole cylinder, $u_l = \frac{3}{4}\frac{\sigma_s r_2^2}{Er_1}$. The ratio of the displacements in the above two cases is $u_l/u_e = (r_2/r_1)^2$.

3.6 Elastoplastic Bending of Beams

The plastic ultimate load of an elastoplastic structure is the maximum value that characterizes the load-carrying capacity of the structure. The structural design based on the plastic ultimate bearing capacity can not only give full play to the plastic properties of the material, but also obtain the parameters reflecting the true safety margin of the structure. In order to determine the plastic limit load of the structure, an elastoplastic analysis method can be used. In such analysis, it is necessary to understand the entire loading process, and since the physical relationship of the material is nonlinear, it is convenient to solve the simple problem only. If the deformation limit process of the structure is not considered, and the plastic limit state is directly analyzed, the analysis of the problem is greatly simplified, and the obtained plastic limit load is exactly the same as that obtained by the elastoplastic analysis method [2, 3].

For the limit analysis of the structure, the following three aspects can be obtained:

(1) The plastic limit load of the structure;
(2) The distribution of stress (or internal force) when the plastic limit state is reached;
(3) A damage mechanism formed at the moment when the structure reaches the plastic limit state.

Under the action of variable load, the analysis of the failure mode and bearing capacity of the structure is the content of structural plasticity stability. In general, the load value of the structure under stability is much lower than the plastic limit load; in some cases, the stability load is close to or equal to the plastic limit load. The following basic assumptions are used in the elastoplastic analysis and the limit analysis of the beam bending:

(1) Only the normal stress is considered in the cross section of the beam, and compressive stress is ignored; since the shear stress component of the plastic zone is zero, only normal stress is included in the yield condition of the beam.
(2) In the case of bending deformation, the cross section of the beam is always kept flat and perpendicular to the deformed beam axis.
(3) Till the moment when the beam reaches the plastic limit state, its deflection is a small amount compared with the cross-sectional dimension, i.e., the assumption of small deflection is still valid before the beam undergoes unconstrained plastic deformation.

3.6.1 Plastic Limit Moment and Plastic Hinge of Beam Section

A simply supported beam with a rectangular cross section (b $\times$ 2 h) is shown in Fig. 3.8, which acts as a concentrated force in the span. It is assumed that the material is ideally plastic. According to hypothesis (1), the stress component is

$$\sigma_x = \sigma_z = \tau_{xy} = \tau_{yz} = \tau_{zx} = 0, \sigma_x = \sigma(x, z). \tag{3.45}$$

The relationship between the normal stress σ and the bending moment M is

$$M = 2b \int_0^h \sigma z \mathrm{d}z. \tag{3.46}$$

From hypothesis (2), it yields

$$\varepsilon_x = \varepsilon(x, z) = -Kz. \tag{3.47}$$

wherein K is the curvature after the beam axis is deflected, and when the deflection w is in the same direction as the z-axis, its curvature is negative. At small deflections, there are

$$K = \mathrm{d}^2 w/\mathrm{d}x^2. \tag{3.48}$$

The yielding condition of the beam section stress is $\sigma = \sigma_s$.

As the load p increases, the middle section of the span undergoes three stress states shown in Fig. 3.9, and the corresponding bending moment is

$$
\begin{aligned}
& M_e = 2bh^2\sigma_s/3, \text{plastic starting status;} \\
& M_s = bh^2\sigma_s[1 - (he/h)^2/3], \text{with an elastic core;} \\
& M_p = bh^2\sigma_s, \text{complete plastic status.}
\end{aligned}
\tag{3.49}
$$

When the section is all plastic, the bending moment M_p is called the plastic limit of the section, and the bending moment can no longer be increased. The plastic

Fig. 3.8 Simply supported beam in bending

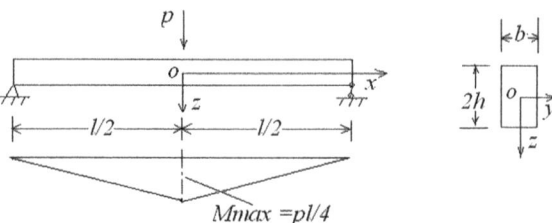

$M max = pl/4$

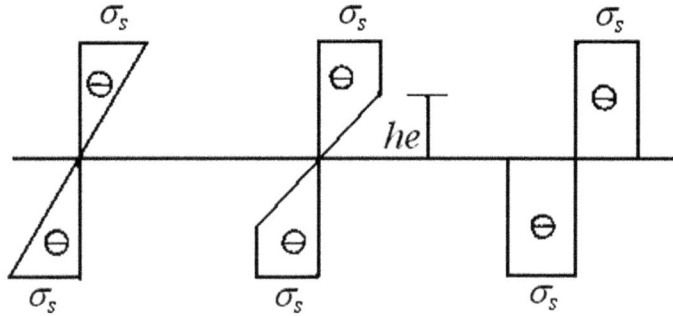

Fig. 3.9 Three stress states

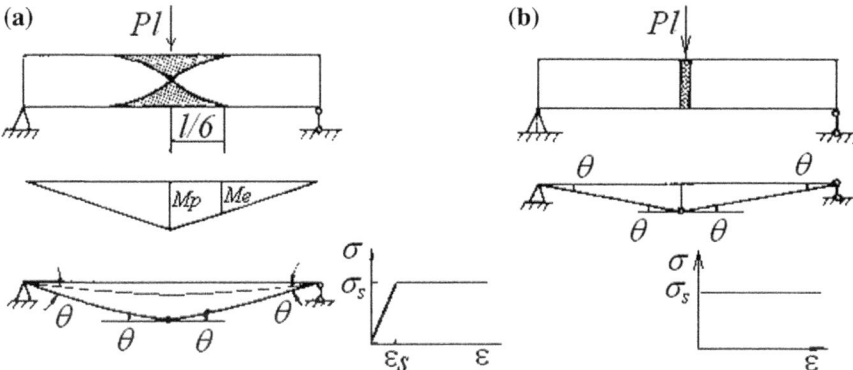

Fig. 3.10 Simply supported beam

zone at this time is shown in the dark part of Fig. 3.10a. Since the upper and lower plastic zones of the mid-span cross section connected each other, the cross sections of the middle and left sides of the span will occur a rotation relatively. Like the function of the ordinary structural hinge, the plastic hinge is identical to the structural hinge in allowing the beam to rotate.

However, the existence of the plastic hinge is due to $M = M_p$ on the section, and the plastic hinge is a one-way hinge; that is, the direction of rotation of the beam section is consistent with the direction of the plastic limit bending moment; otherwise, the plastic hinge will disappear.

If the σ–ε model adopts ideal rigid plasticity, and when the plastic hinge is formed, the plastic zone is limited to a cross section, and the failure mechanism at this time consists of two straight segments (Fig. 3.10b). The plastic hinges in the destruction mechanism are indicated by black dots, while the structural hinges are still drawn as circles.

The ultimate load obtained by the two elastic deformation models of ideal elastoplasticity and rigid phase is the same.

3.6.2 Limit Conditions

The limit condition is also called the generalized yielding condition. It is the condition that the combination of internal stress reaches to the critical value when all the beam sections enter the plasticity. The yield condition, expressed by the stress components, and the assumption of the stress component on the section under the limit state can be used to obtain the limit condition.

Assume that when the section enters the limit state, σ does not change along z, and it yields $M = 2b \int_0^h \sigma z \, dz = bh^2 \sigma$, and considering the yield condition $\sigma = \sigma_s$, the limit condition is $M = M_p$, where M is as shown in the third expression of Eq. (3.49).

For a statically fixed beam (Fig. 3.8), when the middle section of the span exhibits $M = M_p$, i.e., a plastic hinge appears, the beam forms a failure mechanism and loses the ability to continue to carry. In the case of a statically indeterminate beam, it is necessary to form enough plastic hinges to make the beam a destructive mechanism.

3.6.3 Relationship Between Bending Moment and Curvature

When the cross section of the beam is in an elastic state, $\sigma = E\varepsilon$, by using Eq. (3.47), it obtains $K = -\frac{\sigma}{Ez}$. At $z = h$, if $\sigma = \sigma_s$, it can be obtained $K_e = -\frac{\sigma}{Eh} = -\frac{\varepsilon_s}{h}$.

When the deformation of the section is in the elastoplastic state, the stress at $z = he$ reaches σ_s; he is the semi-height of elastic core; it is obtains by using the expressions of K and K_e, $K_s = -\varepsilon_s/he = hK_e/h_e$, or $K_s/K_e = h/he$, where he can be obtained from the second expression of Eq. (3.49); finally, it yields

$$K_s/K_e = \left[3\left(1 - M_s/M_p\right)\right]^{-1/2}. \tag{3.50}$$

If $he = 0$, $M_s = M_p$, $K_s = K_p = \infty$, the section exhibits unconstrained plastic deformation.

In Fig. 3.11, the straight line $0A$ and the curve AB are the M–K relationship curves of the cross section of the simply supported beam. When $K = 3K_e$ and $5K_e$, $M_s = 0.963M_p$ and $0.987M_p$, that is, M is close to M_p.

Fig. 3.11 M–K curve

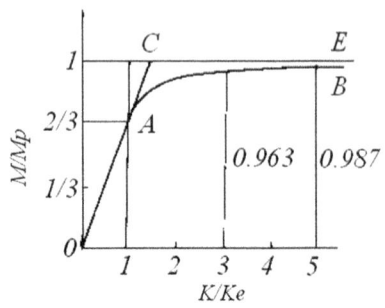

3.7   Stress Analysis of Tube Under Uneven External Load

The geometric compression on casing of oil well under uneven external load is a usual case. Li et al. proposed an expression to solve this problem [4]. Assume that the external load acting on the outer wall of the casing is shown in Fig. 3.12, where p_0 is the pressure generated by the stationary fluid, which is equal at the outer wall of the casing, and its direction is perpendicular to the outer wall of the casing; p_1 is along the x-axis.

The constant load pressure acting on the casing wall is the uneven external load component. Thus, in the Cartesian coordinate system, the stress components corresponding to the external load along all directions can be expressed as follows:

$$p_x = p_1 + p_0 \cos \theta, \quad p_y = p_0 \sin \theta, \tag{3.51}$$

where θ is the angle between the line connecting the point on the outer wall of the casing and the x-axis from the origin.

Convert each component of the stress inside the casing in the Cartesian coordinate system into the polar coordinate system to obtain the expression of the stress component:

$$\sigma_r = -(p_0 + p_1 \cos 2\theta) = -[p_0 + p_1(1 + \cos 2\theta)/2],$$
$$\sigma_\theta = -p_1 \sin 2\theta = -p_1(1 - \cos 2\theta)/2, \tag{3.52}$$
$$\tau_{r\theta} = p1 \sin \theta \cos \theta = p_1 \sin 2\theta/2.$$

From the theory of elastic mechanics [4], the differential equation of the stress on the casing in the polar coordinate system can be obtained

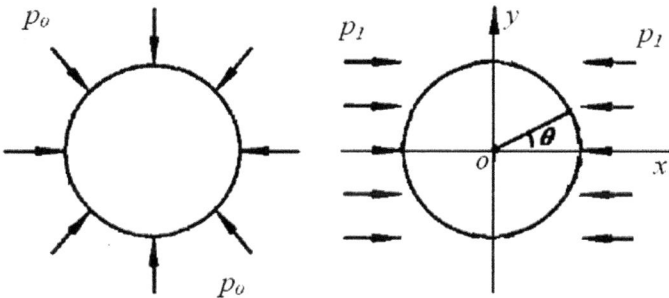

Fig. 3.12 A uneven distribution of external loads on casing

$$\nabla^4 \varphi = \left[\frac{\partial^2}{\partial r^2} + \frac{1}{r} \frac{\partial}{\partial r} + \frac{1}{r^2} \frac{\partial^2}{\partial \theta^2} \right]^2 \varphi = 0, \tag{3.53}$$

where $\varphi(r, \theta)$ is the stress function of the stress applied to the casing.

Let a represent the inner radius of the casing and b express its outer radius. If the casing is also subjected to the internal pressure p_i, take the center of the casing as center of the coordinate system. The stress distribution on the outer wall is

$$\sigma_r|_{r=a} = -p_i, \ \tau_{r\theta}|_{r=a} = 0;$$
$$\sigma_r|_{r=b} = -[p_0 + p_1(1 + \cos 2\theta)/2], \quad \tau_{r\theta}|_{r=b} = p_1 \sin 2\theta/2. \tag{3.54}$$

(1) Stress Analysis

Using the superposition principle, the above problem can be decomposed into a superposition of two boundary value problems, i.e., $\varphi(r, \theta) = \varphi_1(r, \theta) + \varphi_2(r, \theta)$,

$$\nabla^4 \varphi_1 = \left[\frac{\partial^2}{\partial r^2} + \frac{1}{r} \frac{\partial}{\partial r} + \frac{1}{r^2} \frac{\partial^2}{\partial \theta^2} \right]^2 \varphi_1 = 0,$$
$$\sigma_{1r}|_{r=a} = -p_i, \ \tau_{1r\theta}|_{r=a} = 0; \tag{3.55}$$
$$\sigma_{1r}|_{r=b} = -[p_0 + p_1/2], \quad \tau_{1r\theta}|_{r=b} = 0.$$

$$\nabla^4 \varphi_2 = \left[\frac{\partial^2}{\partial r^2} + \frac{1}{r} \frac{\partial}{\partial r} + \frac{1}{r^2} \frac{\partial^2}{\partial \theta^2} \right]^2 \varphi_2 = 0,$$
$$\sigma_{2r}|_{r=a} = 0, \ \tau_{1r\theta}|_{r=a} = 0; \tag{3.56}$$
$$\sigma_{2r}|_{r=b} = -p_1 \cos 2\theta/2, \quad \tau_{2r\theta}|_{r=b} = p_1 \sin 2\theta/2.$$

Solve Eq. (3.55), and it gets

$$\sigma_{1r} = -\left\{ \left[(a/r)^2 - \eta^2 \right] p_i + \left[1 - (a/r)^2 \right] [p_0 + p_1/2] \right\} / (1 - \eta^2),$$
$$\sigma_{1\theta} = \left\{ \left[(a/r)^2 + \eta^2 \right] p_i - \left[1 + (a/r)^2 \right] [p_0 + p_1/2] \right\} / (1 - \eta^2) \tag{3.57}$$
$$\tau_{1r\theta} = 0.$$

where $\eta = a/b$.

As to Eq. (3.56), considering the relation among the boundary condition, stress function, and stress components, assume $\varphi_2(r, \theta)$ be the following form,

$$\varphi_2(r, \theta) = f(r) \cos 2\theta, \tag{3.58}$$

From Eq. (3.56), it yields

$$\nabla^4 f(r) = \left[\frac{\partial^2}{\partial r^2} + \frac{1}{r} \frac{\partial}{\partial r} + \frac{1}{r^2} \frac{\partial^2}{\partial \theta^2} \right]^2 f(r) = 0. \tag{3.59}$$

The solutions of Eq. (3.59) are

$$
\begin{aligned}
f(r) &= Ar^4 + Br^2 + C + D/r^2, \\
\sigma_{2r} &= -\left[2B + 4C/r^2 + 6D/r^4 \right] \cos 2\theta, \\
\sigma_{2\theta} &= \left[12Ar^2 + 2B + 6D/r^4 \right] \cos 2\theta, \\
\tau_{2r\theta} &= \left[6Ar^2 + 2B - 2C/r^2 - 6D/r^4 \right] \sin 2\theta.
\end{aligned} \tag{3.60}
$$

Let $K = (1 - \eta^2)^4$, and the constants A, B, C, and D in Eq. (3.60) can be determined by the boundary condition of Eq. (3.56); it yields

$$
\begin{aligned}
A &= -p_1 \eta^2 (1 - \eta^2)/(2\,Kb^2), \\
B &= p_1(1 + 3\eta^4 - 4\eta^6)/(4\,K), \\
C &= -p_1 a^2 (1 - \eta^6)/(2\,K), \\
D &= p_1 a^4 (1 - \eta^4)/4\,K.
\end{aligned} \tag{3.61}
$$

Thus, the total stress in the casing can be written as,

$$
\begin{aligned}
\sigma_r &= \sigma_{1r} + \sigma_{2r}, \\
\sigma_\theta &= \sigma_{1\theta} + \sigma_{2\theta}, \\
\tau_{r\theta} &= \tau_{1r\theta} + \tau_{2r\theta}.
\end{aligned} \tag{3.62}
$$

(2) *Strength Analysis of Casing Under Uneven Compression*

Under condition of the above-mentioned uneven external pressure, the main way of casing failure is that the material has plastic yielding, which causes irreversible deformation of the casing. The equivalent stress at each point in the casing is

$$\sigma_e = \left\{ \left[(\sigma_r - \sigma_\theta)^2 + (\sigma_z - \sigma_\theta)^2 + (\sigma_r - \sigma_z)^2 + 6\tau_{r\theta}^2 \right]/2 \right\}^{1/2}. \tag{3.63}$$

The plastic judgment for any point in the casing is

$$\sigma_e \leq \sigma_s, \tag{3.64}$$

where σ_s is the yield strength of the casing material.

When the equivalent stress at a point in the casing reaches to the yield strength of the casing material, the casing is plastic yielding and ineffective.

In the case of plane strain, the axial stress of the casing is $\sigma_z = v(\sigma_r + \sigma_\theta)$, for the steel material $v \approx 0.33$. Therefore, the equivalent stress of Eq. (3.63) can be simplified to

$$\sigma_e \approx [0.725(\sigma_r - \sigma_\theta)2 + 3\tau_{r\theta}]^{1/2}. \tag{3.63'}$$

Substituting Eq. (3.63') into Eq. (3.64), it yields

$$[(\sigma_r - \sigma_\theta)^2 + 3\tau_{r\theta}^2/0.725]^{0.5} \leq (1/0.725)^{0.5}\sigma_s = 1.1749\sigma_s. \tag{3.65}$$

For actual pipelines, there is generally $0.85 < \eta = a/b < 0.91$ [4].

For the sake of simplicity, the following is only a force analysis of the case where the casing is not subjected to internal pressure ($p_i = 0$).

Analysis shows that σ_e takes an extreme value at $\theta = 90°$, and its value is [4]

$$\sigma_e|_{\theta=90°} = 0.851 (\alpha - \beta). \tag{3.66}$$

where

$$\begin{aligned}
\alpha &\approx 2[1/(1 - \eta^2) - (r - a)/(b - a)](p_0 + p_1/2), \\
\beta &\approx [(4\eta^2 - 18\eta^4 + 4\eta^6 + 5\eta^8 + 5)(r - a)/(b - a) - 6(1 - \eta^4)]p_1/K,
\end{aligned} \tag{3.67}$$

The maximum occurs at $r = a$; it is

$$\sigma_e|_{\theta=90°,r=a} = 0.851\left[2/\left(1 - \eta^2\right)(1 + \varepsilon/2) + 6\left(1 - \eta^4\right)\varepsilon/K\right]p_0, \tag{3.68}$$

where $\varepsilon = p_1/p_0$.

ε is in fact a parameter characterizing the unevenness of the external load. When $\varepsilon = 0$, it is a uniform external load. The larger ε, the more uneven the external load.

Substituting Eqs. (3.67) and (3.68) into Eq. (3.65), the corresponding external stress condition when the maximum value of the equivalent stress σ_e at $r = a$ and $\theta = 90°$ reaches to the yield condition, that is,

$$\left\{\left[2/\left(1 - \eta^2\right)\right](1 + \varepsilon/2) + 6\left(1 - \eta^4\right)\varepsilon/K\right\}p_0 \leq 1.175\sigma_s. \tag{3.69}$$

The results indicate the application of the above approach [4].

References

1. Kassir M (2018) Applied elasticity and plasticit. CRC Press, Taylor & Francis Group, Boca Raton
2. Xu B, Li X (1995) Applied elastoplastic mechanics. Tsinghua University Press, Beijing

3. Sadd MH (2014) Elasticity theory, applications, and numerics, 3rd edn. Academic Press of Elsevier, Waltham
4. Li J, Zheng M, Zhang J (2007) Influence of non-uniform external-pressure load on the load-bearing capacity of casing. J Xi'an Shiyou Univ (Nat Sci Ed) 22(1):100–105

Chapter 4
Transfer and Mitigation of Stress Concentration at the Root of a Notch for Highly Ductile Materials

Abstract In this chapter, the stress distribution around the notch root of a structure in plastic status is employed to represent the transfer and mitigation of stress concentration in the structure consisting of high ductile material; the elastic–plastic solution for a notched component and the slip-line field approach formulated in plastic mechanics are combined to perform the study.

4.1 Introduction

It is very usual thing for notches and holes appearing in machines and structures, which result in stress concentration inevitably. Stress concentration is an attractive problem to both machine designers and users, which is concerned in structures under tension. The stress distribution in a thin rectangular plate is uniform under condition of a uni-axial tension, Fig. 4.1a. However, as a circular hole is introduced into the plate, it disturbs the uniform distribution of stress near the hole, and results in a significantly higher tensile stress near the hole than average stress, Fig. 4.1b.

In general, the tensile stress near the hole is enhanced higher than the average stress in elastic range, such a phenomenon is called stress concentration, see Fig. 4.1. Theoretically, as the size of the plate is infinite large, the tensile stress near the hole is three times of that of the average stress in elastic range. While, as to a rectangular plate with finite width, the stress concentration factor is a function of the ratio of the hole's radius to the plate width. The problem of the stress concentration is of practical interest due to its effects on fracture and fatigue.

© Springer Nature Singapore Pte Ltd. 2019
M. Zheng et al., *Elastoplastic Behavior of Highly Ductile Materials*,
https://doi.org/10.1007/978-981-15-0906-3_4

Fig. 4.1 Loading specimen

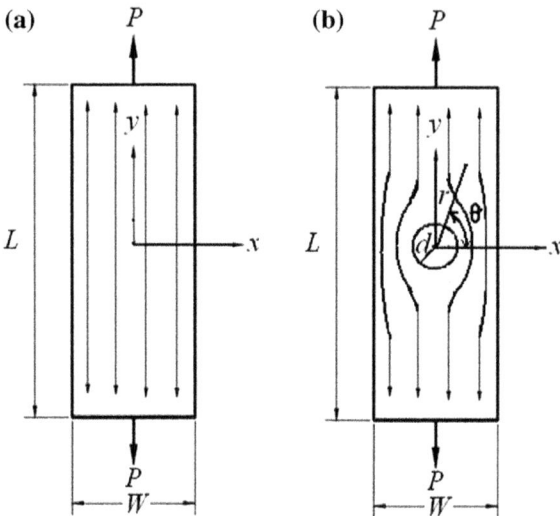

4.2   Stress Concentration at the Edge of the Hole

The actual value of stress concentration at the edge of a hole is quite important for engineering structures. The stress distribution around a hole in an infinite plate under a uniform load was first proposed by Kirsch in elastic range [1], which reads as,

$$
\begin{aligned}
\sigma_r &= \frac{\sigma_a}{2}\left(1 - \frac{a^2}{r^2}\right) + \frac{\sigma_a}{2}\left(1 - \frac{4a^2}{r^2} + \frac{3a^4}{r^4}\right)\cos 2\theta \\
\sigma_\theta &= \frac{\sigma_a}{2}\left(1 + \frac{a^2}{r^2}\right) - \frac{\sigma_a}{2}\left(1 + \frac{3a^4}{r^4}\right)\cos 2\theta \\
\tau_{r\theta} &= -\frac{\sigma_a}{2}\left(1 + \frac{2a^2}{r^2} - \frac{3a^4}{r^4}\right)\sin 2\theta
\end{aligned}
\tag{4.1}
$$

The circumferential normal stress at $\theta = \pi/2$ in the y-axis direction is

$$
\sigma_\theta = \sigma_a\left(1 + \frac{1}{2}\frac{a^2}{r^2} + \frac{3a^4}{2r^4}\right),
\tag{4.2}
$$

in which σ_r, σ_θ, and $\tau_{r\theta}$ are the radial, tangential, and shear-stress components, respectively, σ_a is the remote unidirectional stress and a is the radius of the hole. The *stress concentration factor* is defined here as the *ratio of the maximum stress at the edge of the hole and the remotely nominal uniform stress*,

$$K_t = \sigma_{max}/\sigma_a, \tag{4.3}$$

For an infinite plate with a hole under condition of uniformly remote tension, the stress concentration factor at the edge of a hole is $K_t = 3$.

4.3 Transfer and Mitigation of Stress Concentration in the Root of a Notch for Highly Ductile Materials

Since 1950s, much attention has been paid to study the stress concentration behavior near a notch in structure [2].

The concept of stress concentration was originally put forward in elastic status, which has been extended to correlate the remotely nominal stress and the local stress–strain at the root of a notch if local plastic yielding appears around the notch in structure.

In this chapter, the stress distribution around the notch root of the structure in plastic status is employed to represent transfer and mitigation of stress concentration at the structure consisting of high ductile material; the elastic–plastic solution for a notched component and the slip-line field approach formulated in plastic mechanics are combined to perform the study.

A shallow notch is shown in Fig. 4.2, in which the notch root is approximated by a semi-elliptical shape [3]. Shi et al. once developed a simple method to estimate the maximum normal stress and plastic zone size at a shallow notch [3]. According to their study, the depth of the notch c is much smaller than the width of the specimen in the x-direction to be a shallow notch, and the stresses in the vicinity of the notch can be obtained by using elastic theory. Until plastic yielding, the maximum value of elastic tensile stress, σ_{max}, occurs at the root of the notch, and it can be expressed as [3]

$$\sigma_{max} = \sigma_a \left[1 + 2 \cdot (c/\rho)^{0.5} \right], \tag{4.4}$$

where σ_a is the tensile stress in remote position, and ρ is the curvature radius of notch root.

In the region close to the notch root, the tensile stress along y-direction can be approximated by [3]

$$\sigma_y = \sigma_{max} \cdot [\rho/(\rho + 4r)]^{0.5}, \tag{4.5}$$

where r is the distance from the notch root as shown in Fig. 4.2.

According to plastic yielding criterion, the material element yields as the maximum shear stress or equivalent stress at the notch root approaches the yielding strength of the material, σ_{ys}. Furthermore, a plastic zone forms if the stress increases

Fig. 4.2 Shallow notch
geometry

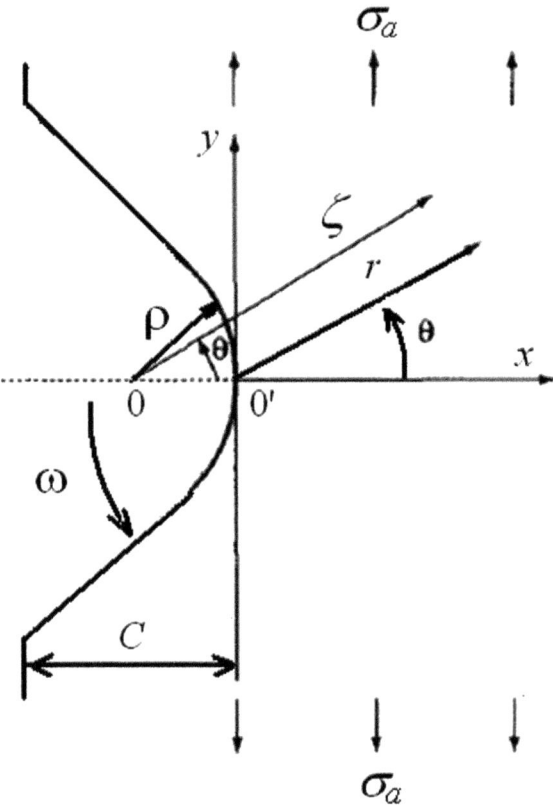

continuously. According to slip-line theory, the tensile stress inside the plastic zone
along y-direction could be expressed as [4]

$$\sigma_{yy} = \sigma_{ys} \cdot [1 + \ln(1 + r/\rho)], \, (r < r_{pz}). \tag{4.6}$$

in which, r_{pz} is the size of the plastic zone.

Equation (4.6) indicates that *the maximum tensile stress along y-direction does
not appear at the root of the notch now, but moves to r = r_{pz}, which indicates that
the concept of stress concentration can only be valid precisely up to the point at
which plastic yielding occurs at the root of a notch.*

According to Shi et al. [3], there is a correlation among the maximum stress
value along y-direction, the plastic size, r_{pz}, the critical plastic zone size, r_β, and the
projected distance, r_o, of the intersection point between the elastic solution and the
slip-line field solution, as shown in Fig. 4.3 (for further details, see [3]).

In addition, a series of shallow notch components with the yielding strength of
540 MPa was analyzed by Shi et al. [3]. According to their results and the general
definition for "stress concentration factor", the actual value of "stress concentration

Fig. 4.3 Stress distribution near notch root after plastic yielding

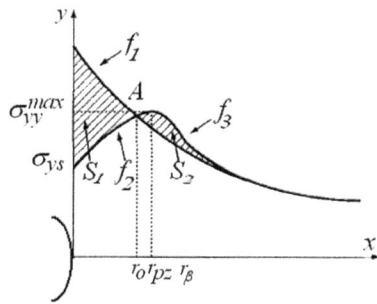

factor" $\kappa \left(\kappa = \sigma_{ys}/\sigma_a \right)$ reduces significantly as long as the plastic yielding zone forms around the notch root, and the ratio of the maximum value of tensile stress σ_{max} at the position r_{pz} along y-direction to the remote tensile stress σ_a, $\chi = \sigma_{max}/\sigma_a$, is really smaller than the theoretical stress concentration factor $K_0 = \left[1 + 2 \cdot (c/\rho)^{0.5} \right]$ for the shallow notch [3], as is shown in Table 4.1.

In order to consider the effect of plastic deformation around notch root, Neuber proposed a rule. In his rule, the product of the local stress and strain concentration factors is supposed to be as the square of the theoretical stress concentration factor [2]. While, Moski and Glinka supposed that the ratio of the local strain energy density at a notch root to that of the remote nominal stress equal to the square of the theoretical stress-concentration factor [5]. Polak conducted the comparison and the predictions of various methods with the experimental data [6]. It was found that significant disagreement occurs for higher loadings [5, 6]. On the other hand, in fracture research for ductile materials, it was also shown that crack initiation does not always occur at the notch root for a tensile notched bar that has enough ductility, instead the crack forms and starts first at the center of the (notched) bar [7, 8]. Obviously, not all these phenomena can be understood in the common view point of stress concentration. This indicates that the common idea of stress concentration at the notch root does not always reflect the real case for a notched component under plastic yielding condition. Therefore, the detailed study of stress concentration around a notch root must be performed by some means, such that a proper understanding can be obtained [9].

Zheng et al. conducted an analysis of the stress concentration factor for a shallow notch by the slip-line field method [9], which declared the transfer and mitigation of stress concentration at the notch root of a structure consisting of high ductile material.

According to the consistency (or conformity) requirement for stress at the interface of elastic and plastic zones, the stress in elastic zone at the interface along y-direction could equal to the stress in plastic zone at the same interface along y-direction [9].

Table 4.1 Comparison of the theoretical stress concentration factor and actual one around the notch root under plastic condition

c (mm)	ρ (mm)	σ_a (MPa)	σ_{max}	r_{pz} (µm)	r_{pz}/ρ σ_{ys}	$K_0 \cdot \sigma_a/$ σ_{ys}	K_0	$\kappa = \sigma_{ys}/\sigma_a$	$\chi = \sigma_{max}/\sigma_a$
0.05	0.05	300	674	14.1	0.282	1.67	3.00	1.80	2.25
0.05	0.05	400	785	28.7	0.574	2.22	3.00	1.35	1.96
0.05	0.05	500	885	44.7	0.894	2.78	3.00	1.08	1.77
0.10	0.05	200	623	8.3	0.166	1.42	3.83	2.70	3.12
0.10	0.05	300	767	26.1	0.522	2.13	3.83	1.80	2.56
0.10	0.05	400	895	46.4	0.928	2.84	3.83	1.35	2.24
0.10	0.05	500	1006	68.5	1.370	3.54	3.83	1.08	2.01
0.20	0.05	200	712	18.8	0.376	1.85	5.00	2.70	3.56
0.20	0.05	300	885	44.7	0.894	2.78	5.00	1.80	2.95
0.20	0.05	400	1029	73.6	1.472	3.70	5.00	1.35	2.57
0.20	0.05	500	1150	105	2.100	4.63	5.00	1.08	2.30
0.40	0.05	100	585	4.3	0.086	1.23	6.66	5.40	5.85
0.40	0.05	200	830	35.5	0.710	2.47	6.66	2.70	4.15
0.40	0.05	300	1028	73.4	1.468	3.70	6.66	1.80	3.43
0.40	0.05	400	1176	115	2.300	4.93	6.66	1.35	2.94
0.80	0.05	100	674	14.1	0.282	1.67	9.00	5.40	6.74
0.80	0.05	150	836	36.5	0.730	2.50	9.00	3.6	5.57
0.80	0.05	185	935	53.9	1.078	3.08	9.00	2.92	5.05
0.80	0.05	200	974	61.7	1.234	3.33	9.00	2.70	4.87
0.80	0.05	300	1176	118	2.360	5.00	9.00	1.80	3.92
1.00	0.05	100	710	18.5	0.370	1.84	9.94	5.40	7.10
1.00	0.05	200	1026	72.9	1.458	3.68	9.94	2.70	5.13
1.00	0.05	250	1147	104	2.080	4.60	9.94	2.16	4.59
1.00	0.18	250	862	147	0.817	2.65	5.71	2.16	3.45
1.00	0.30	250	772	161	0.537	2.15	4.65	2.16	3.09
1.00	0.64	250	665	166	0.259	1.62	3.50	2.16	2.66

$$\sigma_{ey}\left(r_{pz}\right) = \sigma_{yy}\left(r_{pz}\right) = \sigma_{ys} \cdot \left[1 + \ln\left(l + r_{pz}/\rho\right)\right], \qquad (4.7)$$

where $\sigma_{ey}\left(r_{pz}\right)$ is the stress at the interface in elastic zone, and $\sigma_{yy}\left(r_{pz}\right)$ is the stress in plastic zone at the interface. The elastic strain, $\varepsilon_{ey}\left(r_{pz}\right)$, at the interface along y-direction can be approximated as

$$\varepsilon_{ey}\left(r_{pz}\right) = \sigma_{ey}\left(r_{pz}\right)/E = \sigma_{ys} \times \left[1 + \ln\left(l + r_{pz}/\rho\right)\right]/E, \qquad (4.8)$$

in which E is the elastic modulus of the material. According to the continuity condition of displacement at the elastic–plastic interface, the elastic strain $\varepsilon_{ey}\left(r_{pz}\right)$, could equal to the plastic strain $\varepsilon_{yy}\left(r_{pz}\right)$ at the interface

$$\varepsilon_{ey}\left(r_{pz}\right) = \varepsilon_{yy}\left(r_{pz}\right) \tag{4.9}$$

As to a notch with a root radius R_0, it will blunt to a new radius R if the notch root is loaded, and R increases with load and deformation. In a polar coordinate system, its origin coincides with that of the circular arc of the notch root. The polar coordinate of a common point P in the plastic region is ζ, and its instant displacement is $u(\zeta)$, then the strain can be written as

$$\varepsilon_\zeta = du(\zeta)/d\zeta, \quad \varepsilon_\theta = u(\zeta)/\zeta, \tag{4.10}$$

in which θ is the angle in the polar coordinate system.

The continuity condition for plane strain is as follows

$$d\varepsilon_\theta/d\zeta + \left(\varepsilon_\theta - \varepsilon_\zeta\right)/\zeta = 0, \quad \varepsilon_\theta + \varepsilon_\zeta + \varepsilon_z = 0, \quad \varepsilon_z = 0 \tag{4.11}$$

which leads to $d\varepsilon_\theta/d\zeta = -2\,\varepsilon_\theta/\zeta$.

At the notch root, $\zeta = R$, if the radial displacement is ΔR, the radial strain is [8]

$$\varepsilon_\theta|_{\zeta=r} = \Delta R/R. \tag{4.12}$$

Therefore, the strain at a common point P in the plastic region could be derived from Eqs. (4.10) to (4.12), which is

$$\varepsilon_\theta = \Delta R \cdot R/\zeta^2 = -\varepsilon_\zeta. \tag{4.13}$$

Besides, the equivalent strain is

$$\begin{aligned} \varepsilon_{eq} &= \left(2^{0.5}/3\right) \cdot \left[\left(\varepsilon_\theta - \varepsilon_\zeta\right)^2 + \varepsilon_\theta^2 + \varepsilon_\zeta^2\right]^{0.5} \\ &= 2R \cdot \Delta R/\left(3^{0.5}/\zeta^2\right) = 2\varepsilon_\theta^2/3^{0.5}. \end{aligned} \tag{4.14}$$

Let (4.8) equal to (4.13) at $\zeta = \rho + r_{pz}$ (i.e., at the interface of elastic and plastic zones), the following is obtained

$$\Delta R/R = \left(\rho + r_{pz}\right)^2 \cdot \sigma_{ys} \cdot \left[l + \ln\left(1 + r_{pz}/\rho\right)\right] \Big/ \left(E \cdot R^2\right), \tag{4.15}$$

$$\varepsilon_\theta|_{\zeta=r} = \left(\rho + r_{pz}\right)^2 \cdot \sigma_{ys} \cdot \left[1 + \ln\left(1 + r_{pz}/\rho\right)\right] \Big/ \left(E \cdot R^2\right). \tag{4.16}$$

Since R and ρ in fact represent the curvature radius of the same notch, $R = \rho$ in Eq. (4.16), thus it obtains

$$\varepsilon_\theta|_{\zeta=\rho} = \left(\rho + r_{pz}\right)^2 \cdot \sigma_{ys} \cdot \left[1 + \ln\left(1 + r_{pz}/\rho\right)\right] \Big/ \left(E \cdot \rho^2\right). \tag{4.17}$$

Under condition of $\theta = 0$, $\varepsilon_\theta = \varepsilon_{yy}$, $\sigma_\theta = \sigma_{yy}$, according to the concept of the stress concentration factor of Neuber's rule [2, 5, 6], the ratio $\eta = \left\{ \left[(\varepsilon_\theta \cdot \sigma_\theta)|_{\zeta=\rho} \right] \Big/ \left[\sigma_a^2/E \right] \right\}^{0.5}$ is now the "stress concentration factor" at the notch root under plastic yielding condition, which is related to the so-called theoretical stress concentration factor, K_0. For our shallow notch condition, $K_0 = \left[1 + 2 \cdot (c/\rho)^{0.5} \right]$. Further more, in the light of Eq. (4.17), the ratio η becomes,

$$
\eta = \left\{ \left[(\varepsilon_\theta \cdot \sigma_\theta)|_{\zeta=\rho} \right] \Big/ \left[\sigma_a^2/E \right] \right\}^{0.5}
= \left\{ (\rho + r_{pz})^2 \cdot \sigma_{ys}^2 \cdot \left[1 + \ln(1 + r_{pz}/\rho) \right] \Big/ (\sigma_a^2 \rho^2) \right\}^{0.5} \tag{4.18}
$$

Therefore, the validity of the stress concentration factor now becomes an invariability problem of the ratio parameter η [9].

For a given material and geometry of component, the elastic modulus, the yielding strength and the theoretical stress concentration factor K_0 are all well determined, however, ratio parameter η is undetermined according to Eq. (4.18).

The plastic zone size and the maximum stress value were assessed for some notched components by varying loading conditions [3], which is now employed to assess the variation of ratio parameter η with respect to loadings. Table 4.2 shows the calculated results of η by using Eq. (4.18). The results indicate that η decreases with respect to load increasing, and it equals to K_0 only if no plastic yielding zone forms at the notch root. The size of plastic yielding zone r_{pz} increases with loading gradually, and *the maximum value of tensile stress along y-direction moves to $r = r_{pz}$ away from the notch root with load for a material with enough ductility as well.*

As mentioned previously, Moski and Glinka proposed an equivalent energy density method, they assumed that the ratio of the local strain energy density, $W_1 = \sigma_{ys}^2/2E + \sigma_{ys} \cdot \varepsilon_{yyp}(\rho)$, to the remotely tensile loading energy density, $W_0 = \sigma_a^2/2E$, equals to the square of the theoretical stress concentration factor, i.e.,

Table 4.2 Variations of η/K_0, γ/K_0 and r_{pz}/ρ with respect to load

$K_0 \cdot \sigma_a/\sigma_{ys}$	1.000	1.149	1.296	1.586	1.872	2.155	2.711	2.957
r_{pz}/ρ	0.000	0.053	0.113	0.244	0.385	0.536	0.853	1.000
η/K_0	1.000	0.940	0.904	0.866	0.852	0.852	0.869	0.880
γ/K_0	1.000	1.005	1.019	1.049	1.080	1.112	1.173	1.198
τ/K_0	1.000	0.972	0.960	0.953	0.959	0.973	1.009	1.027

Notice In the calculations, $K_0 = [1 + 2 (c/\rho)^{0.5}]$

$$\gamma = \left\{ \left[\sigma_{ys}^2/2E + \sigma_{ys} \cdot \varepsilon_{yyp}(\rho) \right] \Big/ \left[\sigma_a^2/2E \right] \right\}^{0.5},$$

$$\varepsilon_{yyp}(\rho) = \varepsilon_{yy}(\rho) - \sigma_{ys}/E, \tag{4.19}$$

in which, ε_{yyp} (ρ) is the plastic strain at notch root in y-direction. Again, the actual Moski and Glinka's "stress concentration factor", γ, can be computed, if the local stress-strain and the remotely nominal stress are known.

Table 4.2 shows the variations of γ versus σ_a as well. It can be seen from these data that γ increases with respect to loading, and it equals to K_0 only if no plastic yielding zone forms at the notch root.

The stress concentration factor at a notch root is valid till plastic yielding starting at the notch root.

It can be seen from Table 4.2 that the variation tendencies of η and γ are not the same, and η is always less than K_0, while γ is usually greater than K_0. For lower stress amplitude, the relative deviations of γ with respect to K_0 is <5% only if the relative size of plastic zone, r_{pz}/ρ, is not over 0.244. However, the relative deviations of η and γ to K_0 are not too small, even up to 20% as the size of plastic zone approaches the value of the curvature radius of the notch tip. Thus, taking all these results into consideration, one could assume a new "equivalent stress concentration factor" in the viewpoint of pure data analysis numerically, τ, which meets the need of following relationship [9],

$$\tau^2 = \eta \cdot \gamma. \tag{4.20}$$

The calculation results are listed in Table 4.2 as well, obviously good results are obtained.

The data presented in Table 4.2 shows that the relative deviations of τ to K_0 is less than 5%, if the size of plastic zone around notch is limited to not over the value of radius of notch curvature.

4.4   The Role and Significance of Stress Concentration Transfer and Relief at the Root of a Notch

Notch effect assessment plays very important roles during safety designing, which is related to the stress concentrations in engineering components. The study on notch effect of materials is of much significance for assessing the sensitivity of materials to notches, holes, grooves, or other geometrical discontinuities. The effects of notch for different kinds of material have been well summarized by Zheng et al. [10]. Several methods have also been applied to reveal the effects of notch.

Recently, the ellipse criterion was further extended into a universal fracture criterion by Qu and Zhang [11]. It is concluded that this criterion has the ability to give quantitatively interpretation of the critical fracture strength of various materials from ductile crystalline metals to brittle ceramic materials in various stress states including tension, compression, shear, and others. This newly developed fracture criterion may provide us with such an opportunity to theoretically analyze the effects of notch for a wide range of materials. For material with very high strength and ductility, stress can be transferred away from the notch root due to plastic deformation, and material may behave strengthening due to strain hardening, therefore, the notch effect to material with very high strength and ductility might be both "notch strengthening" and "notch toughening" due to strain hardening. However, as to a brittle material, it has no enough ductility, the notch effect might be stress concentration and leads to brittle fracture. Zheng et al. once proposed a notch sensitivity factor K_{bN} to reflect the sensitivity of material to a notch [10], $K_{bN} = \alpha \left(E \sigma_f \varepsilon_f \right)^{0.5} / \sigma_b$, α is a constant related to the loading type, $\alpha = 1$ for plane stress state and 0.64 for plane strain state; Thus, K_{bN} is a material constant depending on the tensile properties of material. The higher the value of K_{bN}, the less notch sensitive of the relevant material. The less notch sensitive of ductile material results from the capacity of plastic deformation at the notch root and the blunting at the crack tip, which mitigates the stress concentration at the root of a notch for highly ductile materials.

References

1. Timoshenko SP, Goodier JN (1970) Theory of elasticity, 3rd edn. McGraw-Hill Book Co., New York, USA, p 90
2. Neuber H (1961) Theory of stress concentration for shear-strained prismatic bodies with arbitrary nonlinear stress-strain law. J Appl Mech 28:544–551
3. Shi SQ, Puls MP (1995) A simple method of estimating the maximum normal stress and plastic zone size at a shallow notch. Int J Press Vessels Piping 63:67–72
4. Kachanov LM (1974) Fundamentals of the theory of plasticity. Mir Publishers, Moscow, USSR
5. Moski K, Glinka G (1981) A method of elastic-plastic stress and strain calculation at a notch root. Mater Sci Eng 50:93–100
6. Polak J (1983) Stress and strain concentration factor evaluation using the equivalent energy concept. Mater Sci Eng 50:195–200
7. Thomason PF (1990) Ductile fracture of metals. Pergamon Press, Oxford, UK, pp 136–143
8. Zheng M, Hu C, Luo ZJ, Zheng X (1994) Damage characterization of SAE1020 and 1045 steel under torsion and compression. Theor Appl Fract Mech 21:91–99
9. Zheng M, Niemi E (1997) Analysis of the stress concentration factor for a shallow notch by the slip-line field method. Int J Fatigue 19:191–194
10. Zheng XL, Wang H, Zheng MS, Wang FH (2008) Notch strength and notch sensitivity of materials. Science Press, Beijing, China
11. Qu RT, Zhang ZF (2013) A universal fracture criterion for high-strength materials. Sci Rep 3:1117(1–6). https://doi.org/10.1038/srep01117

Chapter 5
Elastoplastic Problems in the Manufacturing Process of Bimetallic Composite Tubes

Abstract In this chapter, the analysis of hydro-forming process of bimetallic composite tube and the corresponding expressions for predicting hydro-forming pressure and residual stress in the relevant process are briefly presented; the cone-head size effect on the interfacial bonding strength of bimetallic composite tube produced by drawing method is displayed as well. Plane strain assumption and elastic–perfectly plastic material model are employed in the derivation.

5.1 Introduction

Currently, tube transport is widely used in petroleum, metallurgy, chemical engineering, electric power and civil tube network, water and heating supply systems, the safe operation of the tube system is a significant problem due to the attendant corrosions, and such problem grows with the increase use of tube transport tremendously.

In order to solve the corrosion problem, a new type of tube was developed, which is named as bimetallic composite tube, i.e., a liner tube (inner tube) was introduced inside a base tube (outer tube). The duties of mechanical load and anti-corrosion/anti-abrasion are shared by the base tube and the liner tube material, respectively. The connection of both tubes is tightly through deformation or other technique. Consequently, the bimetallic composite tube integrates the full superiorities of both liner tube and base tube materials, i.e., it is not only with excellent advantages in strength, but also preponderance in anti-corrosion, wear resistance, and other properties. In addition, the total economic cost in production is reduced tremendously in principle, and time for repairing and replacement of the corroded tubes is saving. Thus, the bimetallic composite tube could be as a new product, which has the potential in chemical engineering, petroleum, construction, nuclear industry, medical, food, fire, etc.

There are two catalogs of bimetallic composite tube production, i.e., plastic forming mode and non-plastic forming mode in accordance with the details of manufacture. As to the plastic forming technology for bimetallic composite tube,

© Springer Nature Singapore Pte Ltd. 2019
M. Zheng et al., *Elastoplastic Behavior of Highly Ductile Materials*,
https://doi.org/10.1007/978-981-15-0906-3_5

the inner tube bears plastic deformation locally or totally to attach to the outer tube tightly; The plastic forming mode includes mainly hydro-forming method, mechanical drawing method, rolling method, explosive forming method, etc. Plastic forming is a relatively simple and economic technology which has advantage of high efficiency especially for bimetallic composite tube forming.

The hydro-forming and mechanical drawing methods are the most significant approaches in bimetallic composite tube production process. Therefore, it is of great importance to evaluate the deformation behavior and stress conditions in the forming process in the above two methods, so that suitable and effective design of the production process could be worked out.

However, a relatively complex plastic deforming process is involved in hydraulic bulging and drawing processes of bimetallic composite tube production, i.e., the nonlinearities involved in the forming process come from both physical event and geometric side, beside the complex boundary conditions, which make it difficult to propose a strict theoretical analysis for forming processes of bimetallic composite tube through hydro-forming and mechanical drawing methods [1–3]. Finite element simulation has been employed to perform relevant analysis frequently, and promising results are obtained [4–6]. In 1976, the process of a tube expanding into a hole in sheet was studied by Krips for the manufacture of heat exchanger [7, 8], an analytical solution of residual contact pressure for this matter was proposed, it assumed the materials to be ideal elastic–plastic ones for the tube and the sheet which are with the same elastic modulus; in 1982, Masashi Takemoto developed an ideal elastic–plastic material model with Tresca yielding criterion so as to give the correlation between the contact pressure and the bulging pressure [9]; Yan developed an equation to evaluate the residual contact pressure of heat exchanger by ignoring strain hardening in 1988 [10].

In fact, above formulae for residual contact pressure are proposed specially for heat exchanger in the corresponding expansion process of tube to sheet, which is deferent from that of the practical forming process of bimetallic composite tube. Obviously, the analysis for the plastic forming process of bimetallic composite tube integration is a different thing as a whole. Wang et al. once investigated the hydraulic expansion process of bimetallic composite tube by means of graphic method and proposed an expression to predict the hydro-forming pressure [11]. But their experimental data is not consistent with their prediction entirely. Zheng et al. once developed an analysis for hydro-forming process of bimetallic composite tube, and the corresponding expressions to assess hydro-forming pressure and residual stress, assumption of plane strain and elastic–perfectly plastic material model are employed, and appropriate agreement with available experimental data was obtained [12].

In addition, ultra-high pressure vessel is key equipment in modern industry; the double-layer hot-sleeve cylinders impose stringent requirements on their structure, strength, and materials. The stress status in the double-layer hot-sleeve high-pressure vessel cylinder can be divided into pre-stressed state and synthetic (working) state. In the actual situation, the inner and outer cylinders operate as a whole, and the inner and outer cylinders have a slight overmatching with the contact (fit), which is a typically classic overmatching problem [13]. This kind of problem must be solved under the overmatching state, so as to truly simulate the process and the stress state of the pressure vessel under the simulated working condition.

In this chapter, the stress analysis of the hydro-forming process of bimetallic composite tube, the correlations to connect residual stress and hydro-forming pressure in the forming process, and the cone-head size effect on the interfacial bonding strength of bimetallic composite tube produced by drawing method are introduced; elastic–perfectly plastic material model and plane strain assumption are employed.

5.2   Overmatching Problem

Overmatching or interference fit is originally used in the assemble between the shaft and the hole, which means that the size of shaft is over that of the hole slightly, it thus could cause an elastic pressure between the surfaces of the parts after assembly, thereby obtaining a tight joint.

The structure is simple, coaxial, and can withstand large axial forces, torques, and dynamic loads.

Now, the idea and the method are transplanted to the hydro-forming process of bimetallic composite tube integration.

5.2.1   Stress Analysis of Hydro-Forming Process

According to the approach proposed by Zheng et al. [12], the initial state of the bimetallic composite pipe system is shown in Fig. 5.1. Let a and b express the inner and outer radii of the liner tube, respectively; simultaneously, c and d express the inner and outer radii of the base tube, respectively; the liner tube is with the wall thickness of $t = b - a$. It exists a gap, $\Delta = c - b > 0$, between the base and liner tubes before the integrating process.

Figure 5.2 shows the stressed state during bulging. According to general elastoplastic mechanics principle [12, 13], there is an expansion of the liner tube steadily in response to the action of the inner pressure in the bulging forming process, and the inner tube will contact to the base tube finally.

There is no contact pressure until the liner reaches to the base tube to deform it. As elastic–perfectly plastic material model and conditions of plane strain are applied, by using the plastic yielding criterion of Tresca, the internal pressure corresponding the liner pipe plastic yielding is [12, 13],

Fig. 5.1 Initial state of
bimetallic composite tube

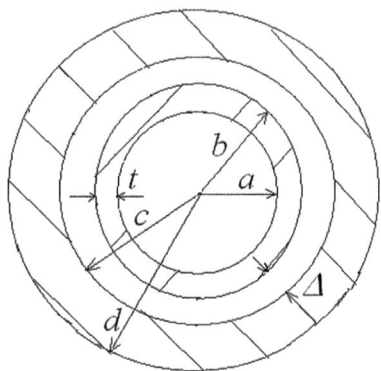

Fig. 5.2 Stressed state of the
bimetallic composite tube
forming process **a** liner tube;
b base tube

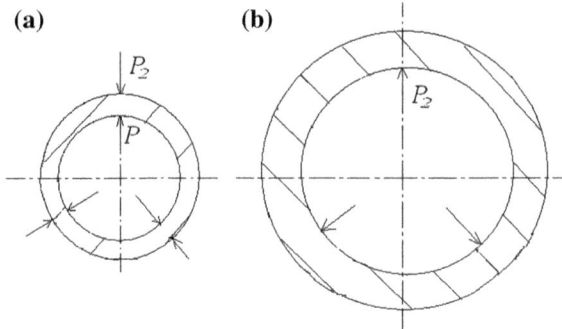

$$P_{10} = \frac{2(b-a)\sigma_{s1}}{b+a} = \frac{2t\sigma_{s1}}{b+a}, \tag{5.1}$$

in which σ_{s1} expresses the plastic yielding strength of the liner material.

If the liner tube expands further, the liner tube will reach to the base tube, and there could be a contact pressure between the base tube and the liner tube. In case of the contact pressure being P_2, the internal pressure inside the liner tube can be written as

$$P = P_{10} + P_2. \tag{5.2}$$

According to elastic–plastic mechanics principle [12, 13], the inner surface of the base tube starts to yield plastically when P_2 arrives at $\frac{\sigma_{s2}}{2}\left(1 - \frac{c^2}{d^2}\right)$. Therefore, the limit value for P_2 that induces the inner surface of the base tube plastic yielding is

$$P_{2c} = \frac{\sigma_{s2}}{2}\left(1 - \frac{c^2}{d^2}\right),$$ (5.3)

here σ_{s2} expresses the yield strength of the base tube material.

At the same time, the internal pressure inside the liner tube P (Eq. 5.2) reaches to its limit P_c, it can be described as:

$$P_c = \frac{2t\sigma_{s1}}{b+a} + \frac{\sigma_{s2}}{2}\left(1 - \frac{c^2}{d^2}\right).$$ (5.4)

In the bulging process, it generates contact pressure as long as the liner tube expands to the base tube, and the base tube starts to withstand the action of the contact pressure P_2. According to elastic–plastic mechanics [12, 13], the radial displacement and the stresses at a point of radius r inside the base pipe are, respectively,

$$u = \frac{c^2 P_2}{E_2(d^2 - c^2)}\left((1 - v_2)r + \frac{(1 + v_2)d^2}{r}\right), \sigma_r = \frac{c^2 P_2}{d^2 - c^2}\left(1 - \frac{d^2}{r^2}\right),$$
$$\sigma_\theta = \frac{c^2 P_2}{d^2 - c^2}\left(1 + \frac{d^2}{r^2}\right).$$ (5.5)

in which, v_2 and E_2 represent Poisson's ratio and elastic modulus of the base tube material.

At this moment, the displacement κ at the inner surface of the base tube is [12, 13]

$$\kappa = u_c = \frac{c^2 P_2}{E_2(d^2 - c^2)} \cdot \left[(1 - v_2)c + \frac{(1 + v_2)d^2}{c}\right].$$ (5.6)

On the other hand, the outer edge of liner pipe expands plastically to the round of the radius $c + \kappa$, which is with the circumference $w = 2\pi(c + \kappa)$.

While, during unloading process, the internal pressure inside the liner tube is absent, the liner tube shrinks back elastically with the outer edge of the circumference of the liner tube drawing back to $w' = 2\pi(c + \kappa) \cdot (1 - \sigma_{s1}/E_1)$, in which E_1 expresses the elastic modulus of the lined tube material. Thus, the corresponding radial displacement of the outer circumference of the liner tube is,

$$\varsigma = (c + \kappa) \cdot \sigma_{s1}/E_1 \approx c \cdot \sigma_{s1}/E_1.$$ (5.7)

Thus, there is an overmatching quantity for the liner and base tubes in the process of unloading procedure, which equals to the difference of the displacement κ at the inner wall of the base tube in bulging process and the radial displacement ς

of liner tube due to its natural drawing back during unloading. Furthermore, the residual pressure at the contact surface in unloading state was obtained [12]

$$P_{\text{rc}} = \left(\frac{\kappa - \varsigma}{c}\right) \bigg/ \left\{ \frac{1}{E_1} \cdot \left[\frac{c^2 + (c - t')^2}{c^2 - (c - t')^2} - v_1 \right] + \frac{1}{E_2} \cdot \left[\frac{d^2 + c^2}{d^2 - c^2} + v_2 \right] \right\}, \quad (5.8)$$

in which, t' stands for the instant wall thickness of the liner tube after expansion, and the condition of $t' \cdot (c - t'/2) = t \cdot (b - t/2)$ holds according to the volume invariable principle, and it results in an approximate relation $t' \approx t \cdot b/c$. Furthermore, Eq. (5.8) becomes [12]

$$P_{\text{rc}} = \left\{ \frac{cP_2}{E_2(d^2 - c^2)} \left[(1 - v_2)c + \frac{(1 + v_2)d^2}{c} \right] - \sigma_{s1}/E_1 \right\} \bigg/ \left\{ \frac{1}{E_1} \right.$$
$$\left. \cdot \left[\frac{c^2 + (c - t \cdot b/c)^2}{c^2 - (c - t \cdot b/c)^2} - v_1 \right] + \frac{1}{E_2} \cdot \left[\frac{d^2 + c^2}{d^2 - c^2} + v_2 \right] \right\}. \quad (5.9)$$

Thus, the correlation between the residual pressure P_{rc} at the contact surface after unloading and the internal pressure P during bulging process could be derived with the help of Eq. (5.2),

$$P = P_{10} + P_2 = \frac{2t\sigma_{s1}}{b + a} + \left\{ P_{\text{rc}} \cdot \left\{ \frac{1}{E_1} \cdot \left[\frac{c^2 + (c - t \cdot b/c)^2}{c^2 - (c - t \cdot b/c)^2} - v_1 \right] \right. \right.$$
$$\left. \left. + \frac{1}{E_2} \cdot \left[\frac{d^2 + c^2}{d^2 - c^2} + v_2 \right] \right\} + \frac{\sigma_{s1}}{E_1} \right\} \cdot \left\{ E_2 \cdot (1 - c^2/d^2)/[(1 - v_2)c^2/d^2 + (1 + v_2)] \right\}.$$
$$(5.10)$$

Furthermore, as the integration process of the base tube and the liner tube is complete, there will be a residual compressive stress after unloading only if $\kappa \geq \varsigma$. It leads to the following demand for P_2 at this moment

$$P_2 \geq P'_{2c} = \sigma_{s1} \cdot E_2 \cdot (d^2 - c^2)/\{E_1 \cdot [(1 - v_2)c^2 + (1 + v_2)d^2]\}. \quad (5.11)$$

Furthermore, it gives the following formula [12]

$$P_{2c} \geq P'_{2c}, \text{i.e.,} \frac{\sigma_{s2}}{\sigma_{s1}} \geq \frac{2E_2}{E_1} \bigg/ [(1 + v_2) + (1 - v_2)c^2/d^2]. \quad (5.12)$$

If the Poisson's ratio and elastic modulus of the base tube are the same as those of the liner tube, Eq. (5.12) reduces to,

$$\sigma_{s2} \geq \sigma_{s1} \bigg/ \left\{ 1 - 0.5 \left[(1 - v)\left(1 - \frac{c^2}{d^2}\right) \right] \right\}. \quad (5.13)$$

Since $c < d$, Eq. (5.13) indicates that there exists a basic condition $\sigma_{s2} > \sigma_{s1}$ for the bulging manufacture of bimetallic composite tube with the liner tube possessing the same elastic modulus and Poisson's ratio as those of the base tube material, and Eq. (5.13) gives the detailed relationship.

In addition, an alternate relationship can be obtained by combing Eqs. (5.1) and (5.2),

$$P_2 = P - P_{10} = P - \frac{2t\sigma_{s1}}{b+a}. \tag{5.14}$$

Substituting Eq. (5.14) into Eq. (5.9), it obtains

$$P_{\text{rc}} = \left\{ \frac{\left(P - \frac{2t\sigma_{s1}}{b+a}\right)}{E_2(1 - c^2/d^2)} \cdot \left[(1 - v_2)c^2/d^2 + (1 + v_2)\right] - \sigma_{s1}/E_1 \right\} \bigg/ \left\{ \frac{1}{E_1} \right. \tag{5.15}$$
$$\left. \cdot \left[\frac{c^2 + (c - t \cdot b/c)^2}{c^2 - (c - t \cdot b/c)^2} - v_1\right] + \frac{1}{E_2} \cdot \left[\frac{d^2 + c^2}{d^2 - c^2} + v_2\right] \right\}.$$

Equation (5.15) expresses the relationship between the residual stress and the internal bulging pressure.

5.2.2 Stress Analysis of Mechanical Drawing Process

Figure 5.3 shows the production process of bimetallic composite tube integration by drawing method.

As to drawing method, the expressions of the contact pressure, internal pressure expanding the inner tube, and the unloading residual pressure P_{rc}, are the same as those of the "hydro-forming process" developed in the previous section [12].

According to Coulomb's friction law, if the sliding friction coefficient between the cone and the liner tube is μ during the drawing process, the friction force equals

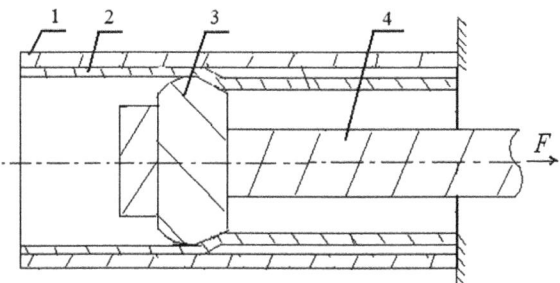

Fig. 5.3 Drawing method for manufacturing bimetallic composite tube

1. Base tube, 2. Liner tube, 3. Cone, 4. Drawing bar

to the product of μ and the bulging force of the cone acting the liner tube, which in turn is also equal to the drawing force F for cone to slide forward within the liner tube, therefore [12],

$$
F = 2\pi c l \mu P = 2\pi c l \mu \cdot \left\{ \frac{2 t \sigma_{s1}}{b+a} + \left\{ P_{rc} \cdot \left\{ \frac{1}{E_1} \cdot \left[\frac{c^2 + (c - t \cdot b/c)^2}{c^2 - (c - t \cdot b/c)^2} - \nu_1 \right] \right. \right. \right.
$$
$$
\left. \left. \left. + \frac{1}{E_2} \cdot \left[\frac{d^2 + c^2}{d^2 - c^2} + \nu_2 \right] \right\} + \frac{\sigma_{s1}}{E_1} \right\} \cdot \left\{ E_2 \cdot \left(1 - c^2/d^2 \right) / \left[(1 - \nu_2) c^2/d^2 + (1 + \nu_2) \right] \right\} \right\}.
$$

$$(5.16)$$

In Eq. (5.16), l represents the contact length of the liner tube and the cone at the maximum cone diameter.

5.3   Application in the Manufacturing Process of Bimetal Composite Tubes

5.3.1   Interface Bonding Strength and Residual Stress

The hydro-forming process in a system called "self-energizing seal" was reported by Wang et al. [11], the liner tube is stainless steel tube with the size of $144 \times 2 \times 3000$ mm, the base tube is carbon steel tube with the size of $159 \times 6 \times 3000$ mm, the gap is $\rho = 1.5$ mm. The basic parameters of materials properties of the tubes are shown in Table 5.1.

Wang reported the variation of the experimental residual contact pressure P_{rc} with respect to the hydro-forming pressure P [11], which is shown in Fig. 5.4 by the cross symbol "×".

The experimental data was employed to check the applicable of Eq. (5.15) [13].

Table 5.1 Basic parameters of tube materials

Name	Material	Yielding strength σ_s (MPa)	Ultimate tensile strength σ_b (MPa)	Elastic modulus E (GPa)	Strain hardening flow stress σ_s' (MPa)	Poisson's ratio
Liner tube	1Cr18Ni9Ti	213	590	198	246[a]	0.3
Base tube	20#	281	480	200		0.3

[a]The flow stress of the liner pipe material at 2.08% strain is $\sigma_{s1}' = 246$ MPa based on the strain–stress curve of material

Fig. 5.4 Comparison of predictions with the experimental data

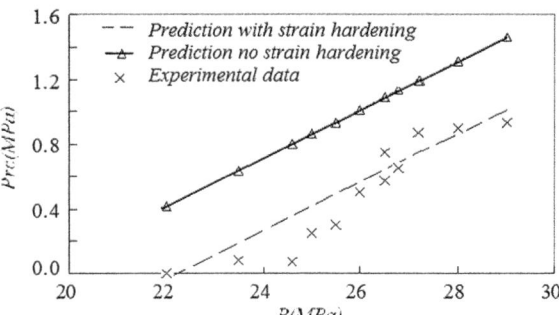

Substituting the basic parameters of the material properties shown in Table 5.1 and yielding strength 213 MPa of the liner tube 1Cr18Ni9Ti into Eq. (5.15) directly, i.e., strain hardening is not considered, it derived an expression [13]

$$P_{rc} = 0.150P - 2.889, \ P \text{ in MPa}. \tag{5.17}$$

The solid triangle line in Fig. 5.4 shows the prediction by Eq. (5.17). As can be seen from Fig. 5.4, the trend of the prediction of Eq. (5.17) agrees with experimental data, however, the detailed result of predicted residual stress is higher than the experimental data.

Moreover, if we consider a certain strain hardening for the liner tube, i.e., taking the strain hardening flow stress 246 MPa of 1Cr18Ni9Ti at 2.08% strain as its plastic yielding strength and the relevant data in Table 5.1 into Eq. (5.15), it yields [12, 13]

$$P_{rc} = 0.150P - 3.336, \ P \text{ in MPa}. \tag{5.18}$$

The dotted line in Fig. 5.4 also shows the prediction of Eq. (5.18). As can be seen from Fig. 5.4, the experimental data is consistent with the prediction of Eq. (5.18) due to the consideration of certain strain hardening of liner material.

5.3.2 Cone Size Effect on the Interfacial Bonding Strength of Bimetallic Composite Tube Integrated by Drawing Method

Zheng et al. also studied the cone size effect on the interfacial bonding strength of bimetallic composite tube integrated by drawing method [14].

It assumes that the interfacial bonding strength Q between the liner tube and the base tube is related to the interfacial residual pressure P_{rc} linearly, i.e., $Q = A \cdot P_{rc} + B$, here,

Table 5.2 Basic material parameters of the both tubes

Name	Material	Elastic modulus E (GPa)	Poisson's ratio	Yield strength σ_s (MPa)	Ultimate tensile strength σ_b (MPa)
Liner tube	0Cr18Ni9Ti	210	0.3	210	520
Base tube	20	210	0.3	281	480

A and B are parameters to be determined, it developed a new formula from Eq. (5.15) [14]

$$Q = \alpha/(c - R) + \beta \tag{5.19}$$

In Eq. (5.19), α and β are parameters which related to the exact materials and the geometrical dimensions of the liner and base tubes; R is the radius of the drawing cone; c is the inner radius of the base tube.

It can be seen from Eq. (5.19) that the interfacial bonding strength Q between the base tube and the liner tube increases in a hyperbolic form with the radius of the drawing cone, R.

The preparation of bimetallic composite tubes by drawing method has been studied [14]. The liner tube is stainless steel tube with size of $\Phi76 \times 2.4$ mm, and the base tube is carbon steel tube with size of $\Phi89.7 \times 5 \times 2000$ mm. The basic properties of the material of tubes are shown in Table 5.2 [14].

Quenched and processed steel 40Cr is employed to produce the cone. Graphite powder is used to lubricate and reduce the contact face between the cone and the inner liner tube. The dimensions of the conical heads were 75.00 mm, 75.20 mm, and 75.36 mm, respectively.

The interfacial bonding strength between the liner and base tubes is tested according to the urban construction standard CJT192-2004. Specific use of the pull-off method is used to conduct the interfacial bonding strength with the tensile rate of 3 mm/min. In each test group, three samples are employed.

Table 5.3 gives the variation of test results of interfacial bonding strength with respect to radius of cone R for the composite pipe.

Furthermore, the test results of Table 5.3 are drawn in Fig. 5.5, and the correlation between the interfacial bond strength Q and the cone size parameter R can be obtained.

Table 5.3 Test results of interface bonding strength of composite pipe produced by drawing method

Cone diameter (mm)	75.00	75.20	75.36
Average bond strength (MPa)	0.39	0.45	0.61

Fig. 5.5 Variation of
interfacial bond strength
Q with respect to R

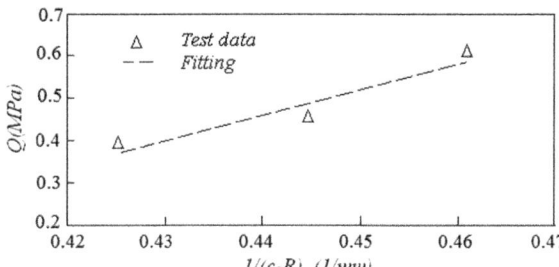

From the experimental results shown in Fig. 5.5, the correlation between the interfacial bonding strength Q and the cone size parameter R was fitted as [14],

$$Q = 6.0856/(c - R) - 2.2171 \qquad (5.20)$$

The unit of c and R in Eq. (5.20) is in mm, and the unit of Q is in MPa. As can be seen from Eq. (5.20), the interfacial bond strength Q and the cone size parameter R have a hyperbolic form.

5.4 Summary

The basic assumptions are plane strain condition and elastic–perfectly plastic material model; the hydraulic bulge forming process of bimetallic composite tube and the relevant formulas are presented; the cone-head size effect on the interfacial bonding strength of bimetallic composite tube produced by drawing method is displayed as well; the comparisons of the theoretical predictions with experimental data indicate the reasonability of the proposed approaches.

References

1. Haghighat H, Asgari GR (2011) A generalized spherical velocity field for bi-metallic tube extrusion through dies of any shape. Int J Mech Sci 53:248–253
2. Marko K, Mohammad J, Yannis PK, Irene JB (2014) Material-based design of the extrusion of bimetallic tubes. Comput Mater Sci 95:63–73
3. Chitkara NR, Aleem A (2001) Extrusion of axi-symmetric bi-metallic tubes from solid circular billets: application of a generalized upper bound analysis and some experiments. Int J Mech Sci 43:2833–2856
4. Lapovok R, Pang Ng H, Tomus D, Estrin Y (2012) Bimetallic copper–aluminium tube by severe plastic deformation. Scripta Mater 66:1081–1084
5. Yang X, Sun F, Zhang Z, Shen H, Guo S (2008) Optimization of drawing parameters' for copper tubes with hollow sinking based on FEM simulation. Chin J Nonferrous Metals 18:2245–2252

6. Xue L, He Y, Liu R, Dai C, Chen J (2005) FEA on empty—sunken steel tube based on ANSYS/LSDYNA. J Plast Eng 12(5):74–77
7. Vedeld K, Osnes Olav Fyrileiv H (2012) Analytical expressions for stress distributions in lined pipes: axial stress and contact pressure interaction. Mar Struct 26:1–26
8. Krips H, Podhorsky M (1976) Hyraolic expansion–a new method for anchoring of tubes. VGB KRAFWERKSTECHINK 56(7):144–153
9. Takemoto M (1984) Tubular Heat Exchanger strength—hydraulic expanding fitting pull-off force detained. Press Vessel 2:68–75
10. Yan H, Yu C (2001) Drawing algorithm for residual contact pressure in heat exchangers during hydraulic expanding. Chem Mach 28:211–214
11. Wang X, Li P, Wang R (2005) Study on hydro-forming technology of manufacturing bimetallic CRA-lined pipe. Int J Mach Tools Manuf 45:373–378
12. Zheng M, Gao H, Teng H, Hu J, Tian Z, Zhao Y (2017) A simplified approach for the hydro-forming process of bi-metallic composite pipe. Arch Metall Mater 62(2):879–883
13. Xu B, Liu X (1995) Applied elastic-plastic mechanics. Tsinghua University Press, Beijing, China, pp 128–233
14. Zheng M, Zhao T, Gao H, Teng H, Hu J (2018) Effect of cone size on the bonding strength of bimetallic composite pipes produced by drawing approach. Arch Metall Mater 63(1):451–456

Chapter 6
Plastic Bending and Failure of Highly Ductile Tubes

Abstract Buckling failure of plastic bending tube is an inevitable phenomenon in practical engineering. Various approaches to deal with buckling failure of plastic bending tube are presented. The experimental data and the rational analysis indicate the model of ellipse approach to describe the ovalization of cross section of bending tube with a rigid–perfectly plastic material mode provides an appropriate estimation for the critical buckling strain. Furthermore, the strain hardening effect on the critical buckling strain of plastic bending tube is presented with comparison with available experimental data as well.

6.1 Introduction

Tube bending is an inevitable phenomenon in practical engineering, which may even lead to tube failure. Tube bending is involved in some production process or practical servicing period of tube. The bent tube or elbow is produced by a set of bending dies to manufacture the product; tube bending process can be divided into two types: cold heading and hot pushing. Bent tube or elbow is used widely in machinery and equipment. When a tube is bent, the outer side of the tube material is pulled, and the inner side position is in compression. In order to get a better surface quality of a bent tube, some measurements are taken, including using advanced benders, using higher-strength molds, or using lubrication products, as well as optimizing bending process.

In recent years, tube transportation is the most economical and reasonable mode for natural gas and oil transport. In practice, there exist multiple seismic and geological disasters, so tube has to bear larger displacement and strain during servicing process inevitably. The tube failure is in whole or in part by displacement or strain.

As to a tube, initiation of bending buckling is employed to define its failure commonly. For a bending tube, the shape of cross section changes with the progress of bending in general. When the tube bending and the cross-sectional shape changing exceed certain degrees, the bending load or moment could no longer increase or even reduces abruptly, which is called buckling initiation of the bending tube.

© Springer Nature Singapore Pte Ltd. 2019
M. Zheng et al., *Elastoplastic Behavior of Highly Ductile Materials*,
https://doi.org/10.1007/978-981-15-0906-3_6

The instability phenomenon for such bending tubes in considering their cross-sectional ovalization was first investigated by Brazier [1]. His result showed that the longitudinal tension and compression at the outer and inner sides of an initial straight pipe uniformly bending arrest the applied bending moment, it tends to ovalise or flatten the cross-section further as bending progress, and it in turn reduces the flexural stiffness of the bending pipe. Brazier realized that there is a maximum value for flexural stiffness, which can be used to define the critical moment at instability. Thereafter, a lot of research work on the analysis of circular tube bending stability has been done, such as elastic instability problems [2–4], elastoplastic instability problems [5], and experimental or analytical analysis [6–11].

In 2006, a variational model was developed by Khurram Wadee et al. to formulate the local deformation of thin-walled elastic tube bending at buckling [12]. Some test data was employed to check the model; however, the model itself is an elastic one. The theoretical characteristics of the initial post-buckling and elastoplastic buckling of cylinders and plates in uniform compression were investigated by Philippe Le Grognec and Anh Le van in 2009 [13]. The analysis was based on the assumption of the J_2 flow theory and a linear isotropic hardening as well as 3D plastic bifurcation approach. The proposed method provided the buckling modes and the initial slope in the bifurcated branch of rectangular plates in case of uniaxial or biaxial compression (-tension) and axial compression of cylinder with various boundary conditions. Poonaya S., Teeboonma U., and Thinvongpituk C. analyzed the plastic collapse of thin-walled circular tubes under bending [14]. 3D geometrical collapse model was employed to analyze the oblique hinge lines in the direction of the longitudinal pipe inside the plastically deformed zone. Furthermore, it evaluated the internal energy dissipation rate of each hinge line. In the derivation of deformation energy rate, perfect plastic material model and in extensional deformation were assumed. A new method to demonstrate the analysis for cross section in the context of the generalized beam theory was developed by Ranzi and Luongo proposed in 2011, [15]. It formulated the problem within the framework of semi-variational method of Kantorovich, and it aimed to describe the linear–elastic behavior of thin-walled members only. In 2012, the effect of variable wall thickness and ovality on collapse loads of tube was investigated by T Christo Michael et al. in case of in-plane bending closing moment [16]. Elastic–perfectly plastic material model and finite element limit analysis were employed in their analysis. The collapse moments were obtained from load–deflection curves. It realized that the ovality has more sensitive influence on collapse loads of the tube bending.

Recently, the critical strain of a pipe at plastic bending buckling is considered as an important index in pipeline design [17–19]. However, the critical strain assessment at buckling, either the regressive formula of experimental data or the classical analytical expression, is questionable due to the shortage of real physical meaning and low accuracy. So, it is hard to meet the practical engineering application [18, 19].

Above analysis shows that it needs a reasonable and rational assessment of tube bending critical buckling strain for practical engineering application.

6.2   Classical (Elastic) Solution for Tube Bending

The elastic equation for circular tube reads

$$\frac{D}{t}\nabla^4\nabla^4 w + \frac{E}{R^2}\frac{\partial^4 w}{\partial x^4} + \sigma\nabla^4\left(\frac{\partial^2 w}{\partial x^2}\right) = 0, \tag{6.1}$$

in which, R, t, and w are the radius, thickness, and deflection of the circular tube, respectively; E is elasticity modulus of the circular tube; $D = Et^3/[12(1-v^2)]$ is the flexural rigidity of the tube, and σ is the uniform stress on the circular tube along its axial direction.

The classical (elastic) solution of axial minimum uniform stress σ at elastic buckling for elastic equation of circular tube is [20]

$$\sigma_{cr} = p_{cr} = \frac{E}{\sqrt{3(1-v^2)}}\frac{t}{r}. \tag{6.2}$$

in which, r is the radius of the circular tube.

For common metallic material, such as steel, its Poisson's ratio $v = 0.3$, while for aluminum and copper, their Poisson's ratio $v = 0.34$ [21], Eq. (6.2) reduces

$$\sigma_{cr} = 0.605E\frac{t}{r}, \quad \text{for steel} \tag{6.3}$$

$$\sigma_{cr} = 0.614E\frac{t}{r}, \quad \text{for aluminium and copper} \tag{6.3'}$$

The corresponding strain is

$$\varepsilon_{cr} = 0.605\frac{t}{r}, \quad \text{for steel} \tag{6.4}$$

$$\varepsilon_{cr} = 0.614\frac{t}{r}, \quad \text{for aluminium and copper} \tag{6.4'}$$

For a bending circular tube, it gives the same result as Eq. (6.4) by assuming that the cross section remains circular and the compressive stress equals to the stress level that corresponds to the classical buckling of a tube in case of axial compression.

The experimental data shows that Eqs. (6.4) and (6.4') overestimate the test results significantly [22, 23].

In general, for a practical tube, the radius–thickness ratio r/t, is about 30–50, the corresponding buckling strain predicted from Eq. (6.4) is $\varepsilon_{cr} = 1.21–2.02\%$, which in fact exceeds the elastic limit strain of the steel tube obviously, says, about 0.2% [22, 23].

It indicates that the prediction of Eq. (6.4) exceeds its scope of application seriously, and it is not valid under actual condition in principle.

6.3   Elastic Buckling During Bending of the Tube Considering Cross-Sectional Change

Li et al. considered the static bending of a thin-walled circular tube, and the cross-sectional change was Brazier type [4].

The longitudinal compressive and tensile stresses make the cross-sectional ovalization, as is shown in Fig. 6.1a. According to the Brazier's assumption, the cross-sectional change is typical elliptical shaped and can be expressed as [4],

$$u = R\xi \cos(2\theta) \tag{6.5}$$

in which, R is the average radius of the original tube, u is the radial displacement of tube, θ is the angle of a point in polar coordinates, and ξ is a dimensionless factor that characterizes the radial displacement, see Fig. 6.1.

Li further assumed that the cross-sectional deformation occurs within its own plane section. Thus, it obtained the total potential energy per unit length [4],

$$U(\xi, C) = \frac{1}{2}\pi EtR^3 \left(1 - \frac{3}{2}\xi + \frac{5}{8}\xi^2\right)C^2 + \frac{3}{8}\frac{\pi E}{1 - v^2}\frac{t^3}{R}\xi^2 - MC \tag{6.6}$$

in which, t is the thickness of the tube wall, v is Poisson's ratio, E is the elastic modulus of tube material, M is the instantaneous bending moment, and C is the longitudinal curvature of the bending tube. The first term on the right side of Eq. (6.6) expresses the potential energy for longitudinal bending, and the second term on the right side of Eq. (6.6) shows elliptical potential energy of the

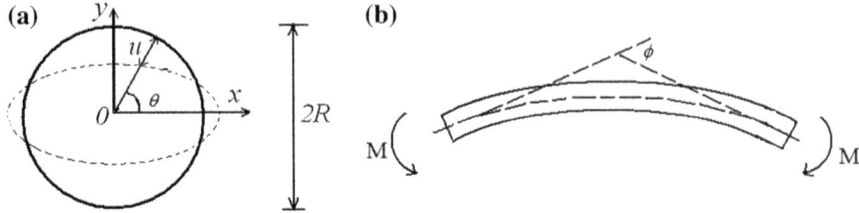

Fig. 6.1 Elliptical cross-sectional model for tube bending

cross-sectional ovalization; the third term on the right side of Eq. (6.6) is the potential energy of the external loading. According to the minimum potential energy principle, it yields [4]

$$\frac{\partial U}{\partial C} = 0, \quad \frac{\partial U}{\partial \xi} = 0 \tag{6.7}$$

And the critical condition for instability is [4],

$$\frac{\partial^2 U}{\partial c^2} \cdot \frac{\partial^2 U}{\partial \xi^2} - \frac{\partial^2 U}{\partial c \partial \xi} \cdot \frac{\partial^2 U}{\partial \xi \partial c} = 0 \tag{6.8}$$

Thereafter, it derives the deformation parameters of a pipe and its critical static moment at instable state as following [4],

$$M_c = 0.388 \frac{\pi E R t^2}{\sqrt{1 - v^2}}, \tag{6.9}$$

$$\xi_c = 0.370, \tag{6.10}$$

$$C_c = 0.731 \frac{t}{R^2} \frac{1}{\sqrt{1 - v^2}}. \tag{6.11}$$

Correspondingly, it yielded the critical strain of the outer fiber line for the bending tube at buckling, which is [22]

$$\varepsilon_c = R(1 - \xi_c) \cdot C_c = 0.630 R \cdot 0.731 \frac{t}{R^2} \frac{1}{\sqrt{1 - v^2}} = 0.461 \frac{t}{R} \frac{1}{\sqrt{1 - v^2}}. \tag{6.12}$$

In Eq. (6.12), though the numerical data $0.461/\sqrt{1 - v^2}$ is less than the result of the classical elastic analytical solution, $\varepsilon_c = 0.605 \frac{t}{R}$, i.e., 0.605, and it is still much higher than the experimental values and an elastic solution [18, 19, 22]. Anyhow, it is a result in considering ovalization of bending pipe cross section.

Moreover, if one neglects the ξ^2 term in Eqs. (6.6) and (6.9) through (6.11) will reduce to the same form as those given by Brazier originally, i.e.,

$$M_c' = \frac{2\sqrt{2}}{9} \frac{\pi E R t^2}{\sqrt{1 - v^2}} = 0.314 \frac{\pi E R t^2}{\sqrt{1 - v^2}}, \tag{6.13}$$

$$\xi_c' = 2/9 = 0.222, \tag{6.14}$$

$$C_c' = \frac{\sqrt{2}}{3} \frac{t}{R^2} \frac{1}{\sqrt{1 - v^2}} = 0.471 \frac{t}{R^2} \frac{1}{\sqrt{1 - v^2}}. \tag{6.15}$$

Similarly, it could yield the critical strain of the outer fiber line for the bending tube at buckling, which can be written as,

$$\varepsilon_c' = R\left(1 - \xi_c'\right) \cdot C_c' = 0.778R \cdot 0.471 \frac{t}{R^2} \frac{1}{\sqrt{1 - v^2}} = 0.366 \frac{t}{R} \frac{1}{\sqrt{1 - v^2}}. \quad (6.16)$$

Again, in Eq. (6.16), though the numerical data $0.366/\sqrt{1 - v^2}$ is less than the result of the classical elastic analytical solution, $\varepsilon_c = 0.605t/R$, i.e., 0.605, it is still much higher than the experimental values [18, 19, 22].

Above discussion indicates that elastic solutions for the tube bending problem at elastic buckling cannot coincide with the actual plastic bending buckling failure of a tube.

6.4 Plastic Buckling During Bending of the Tube

6.4.1 Flattening Cross-Sectional Model for Plastic Bending Tube

A flattening cross-sectional model was developed by Tomasz Wierzbicki et al. to analyze the cross-sectional shape change of tube [24], shown in Fig. 6.2, which assumes a flattening part instead of elliptical cross section.

It is easy to yield the geometric relationship from Fig. 6.2a [24],

$$2b + 2\pi r = 2\pi R, \quad b = \frac{\pi}{2}\delta, \quad (6.17)$$

$$\delta = 2a - 2r, \quad r = R - \frac{\delta}{2}. \quad (6.18)$$

Moreover, it obtained the following expression for the bending moment,

$$\frac{M}{M_0} = \frac{1}{\sqrt{2}}\left(1 - \bar{\delta}\right)\left(1.43 + 1.71\bar{\delta}\right), \quad (6.19)$$

where $M_0 = 4\sigma_0 R^2 t$ is the plastic bending moment of a tube in absolute circular cross-sectional condition with rigid–perfectly plastic mode, M is the applied moment, R is the averaged initial radius of the round tube, σ_0 is the flow stress of

Fig. 6.2 Flattening cross-sectional model for tube bending [24]

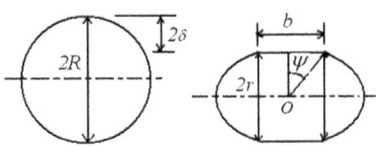

the tube material, t is the wall thickness of tube, and $\bar{\delta} = \delta/(2R)$ is a dimensionless parameter that reflects shape change of cross section.

From Eq. (6.19), the critical dimensionless deformation parameters $\bar{\delta}_c$ at tube bending instability can be obtained,

$$\partial\left(\frac{M}{M_0}\right)/\partial\bar{\delta} = 0, \quad \rightarrow \bar{\delta}_c = 0.0819. \tag{6.20}$$

In addition, it gave an approximate relationship [24],

$$\bar{\delta} = 0.533a^2C/t, \tag{6.21}$$

where C is the longitudinal curvature of the bending tube.

Combing Eqs. (6.20) and (6.21), it obtains the critical value of longitudinal curvature of the tube bending at instability status,

$$C_c = 0.154t/R^2. \tag{6.22}$$

At the same time, it yielded the critical buckling strain of the outer fiber line for the tube bending,

$$\varepsilon_c = (R - \delta_c/2) \cdot C_c = 0.918R \cdot 0.154\frac{t}{R^2} = 0.141\frac{t}{R} \tag{6.23}$$

The parameter in Eq. (6.23), 0.141, is much smaller than the classical elastic solution, i.e., 0.605, and it is smaller than the experimental value as well [18, 19, 23].

As is well known, when a bending load is applied to a tube, the cross-sectional shape of the tube changes more or less. According to the previous analysis, Li took the elliptical cross-sectional shape and the elastic energy approximation for the bending tube [4], a critical strain formula is obtained, i.e., $\varepsilon_c = 0.461t/\left[(1 - v^2)^{1/2} \cdot R\right]$, although the numerical data $0.461/(1 - v^2)^{1/2}$ is smaller than the result of the classical elastic solution, i.e., 0.605, it is still much higher than the experimental values [20]. While in Tomasz Wierzbicki's approach [24], the plastic strain energy was used, and a flattening shape model is employed to approximate the cross-sectional changing of the bending tube; it also leads to a critical strain formula, i.e., $\varepsilon_c = 0.141t/R$, the numerical data here is 0.141, which is much smaller than the result of the classical elastic solution, i.e., 0.605, and it is even smaller than the experimental values [18, 19], which indicates that the approach oversimplified the actual ovalization of tube bending.

6.4.2 *Elliptical Shape Cross Section-Based Expression for Plastic Bending Tube at Buckling*

Ji et al. once proposed a standard ellipse approach to describe the ovalization of cross section of bending pipe so as to deal with the oversimplification problem of Tomasz Wierzbicki's approach for pipe bending [22], and a rigid–perfectly plastic material model was employed in the analysis. The ovalized tube bending and the energy rates of cross-sectional ovalizing were established firstly. Furthermore, these energy rates were combined to perform the analysis of tube bending buckling strain.

(1) *Strain Energy Rate Corresponding to Tube Bending*

According to Brazier effect [1], the cross section of a circular tube steadily becomes elliptical one due to bending, as shown in Fig. 6.1. The ovalization is affected by many factors, among which the geometric size of tube is of great significance. Local buckling appears when the bending moment arrives at a critical value, and there-after, bending moment declines and the bending instability occurs.

For a rigid–perfectly plasticity tube with the original radius R, length l, and wall thickness t, when bending load is applied and it gets into fully plastic state, the pure bending moment corresponding to cross-sectional ovalization could be written as [22],

$$M_p = \frac{4\sigma_{ll}}{3} \left[b^2 a - b_i^2 a_i \right] \tag{6.24}$$

in which, a and b are the semi-lengths of external longer axis and shorter axis; a_i and b_i are the semi-lengths of internal longer axis and shorter axis; and $a = (a_i + t), b = (b_i + t)$. σ_{ll} is a tensile stress along the tube.

For a thin-walled circular tube, $t \ll R$, assume it change into a standard ellipse in response to bending, by introducing a dimensionless parameter, γ, the ellipse can be characterized by $a = R(1 + \gamma)$ and $b = R(1 - \gamma)$. Therefore, by neglecting higher terms of t/R, Eq. (6.24) becomes

$$M_p = 4R^2 t \sigma_{ll} \left[1 - \frac{2}{3}\gamma - \frac{1}{3}\gamma^2 \right] = M_0 \left[1 - \frac{2}{3}\gamma - \frac{1}{3}\gamma^2 \right] \left(\frac{\sigma_{ll}}{\sigma_0} \right). \tag{6.24'}$$

In Eq. (6.24'), σ_0 is the flow stress of the tube material, $M_0 = 4\sigma_0 R^2 t$ is bending moment of a perfect round tube in perfectly plastic state.

Furthermore, the bending strain energy rate of the tube is [22]

$$\dot{W}_b = M_p \dot{\phi} = M_0 \left[\left(1 - \frac{2}{3}\gamma - \frac{1}{3}\gamma^2 \right) \right] \cdot \left(\frac{\sigma_{ll}}{\sigma_0} \right) \dot{\phi}. \tag{6.25}$$

$\dot{\phi}$ is the changing rate of the bending angle of the pipe subjected bending load in Fig. 6.1b.

Geometrically, there is a correlation between bending angle ϕ and the longitudinal curvature C of the tube with length l, $\phi = lC$, so, $\dot{\phi} = l\dot{C}$. Thus, Eq. (6.25) can be further rewritten as,

$$\dot{W}_b = M_p \dot{\phi} = M_0 \left[\left(1 - \frac{2}{3}\gamma - \frac{1}{3}\gamma^2 \right) \right] \cdot \left(\frac{\sigma_{ll}}{\sigma_0} \right) l\dot{C}. \tag{6.26}$$

(2) *Strain Energy Rate Corresponding to the Cross-Sectional Ovalization of a Bending Tube*

For a tube with the thickness t and length l, the cross-sectional ovalization can be described as follows.

According to Tomasz Wierzbicki [24], for an elliptical tube, the strain energy rate due to the cross-sectional ovalizing in the fully plastic state is [22]

$$\dot{W}_{\text{ovalization}} = \oint_S \dot{K}_{\theta\theta} \cdot M_{\theta\theta} \cdot \mathrm{d}s, \tag{6.27}$$

in which, $M_{\theta\theta} = \sigma_{\theta\theta}(t^2 l/4)$ is the plastic moment corresponding to the elliptical arc in Fig. 6.1a; $\dot{K}_{\theta\theta}$ is the changing rate of the elliptical arc curvature; and $\mathrm{d}s$ is the elliptical arc length.

Mathematically, in rectangular coordinate system, each point on the elliptical arc can be reflected by following equation,

$$x = a \cos\theta = R(1 + \gamma)\cos\theta, y = b\sin\theta = R(1 - \gamma)\sin\theta. \tag{6.28}$$

The curvature in the elliptical arc can be derived as,

$$K_{\theta\theta} = \frac{1}{R} \frac{(1 - \gamma^2)}{\left[(1 - \gamma)^2 + 4\gamma\sin\theta \right]^{3/2}} \tag{6.29}$$

There exists a critical angle θ_c, at which $K_{\theta\theta c} = 1/R$. From Eq. (6.29), it yields

$$\theta_c = \arcsin\left\{ \frac{\left[(1 - \gamma^2)^{1/3} + (1 - \gamma) \right] \cdot \left[(1 - \gamma^2)^{1/3} - (1 - \gamma) \right]}{4\gamma} \right\}^{1/2} \tag{6.30}$$

Equation 6.30 gives the variation of θ_c with respect to the parameter γ, and it shows that θ_c decreases rapidly as soon as γ reaches to 0.97, otherwise θ_c decreases with γ slowly.

From Eq. (6.29), it can be seen, for $\theta < \theta_c$, $K_{\theta\theta} > K_{\theta\theta c}$, and for $\theta > \theta_c$, $K_{\theta\theta} < K_{\theta\theta c}$.

Especially, at $\theta = 0$ and $\pi/2$, the value of the curvature K is, respectively,

$$K_{\theta\theta}|_{\theta=0} = \frac{1}{R} \frac{(1+\gamma)}{\left[(1-\gamma)^2\right]}, \quad K_{\theta\theta}|_{\theta=\pi/2} = \frac{1}{R} \frac{(1-\gamma)}{\left[(1+\gamma)^2\right]}. \tag{6.31}$$

Furthermore, through lengthy and complicated analysis, the integral Eq. (6.27) was approximately written as [22],

$$\dot{W}_{\text{ovalization}} = \frac{\pi t l}{32R^2} \cdot M_0 \cdot \left[\frac{(3+\gamma)}{(1-\gamma)^3} \cdot (1 - 0.95\gamma)^{0.5} \right.$$
$$\left. - \frac{(3-\gamma)}{(1+\gamma)^3} \cdot \left[2 - (1 - 0.95\gamma)^{0.5} \right] \right] \cdot \left(\frac{\sigma_{\theta\theta}}{\sigma_0} \right) \cdot \dot{\gamma} \tag{6.32}$$

(3) Plastic Yielding Condition and Total Strain Energy Rate of a Bending Tube

Refer to Tomasz Wierzbicki's method [24], the plastic yielding condition of bending tube problem can be written as follows,

$$\left(\frac{M_\theta}{M_0} \right)^2 + \left(\frac{N_{ll}}{N_0} \right)^2 = 1. \tag{6.33}$$

And furthermore, it follows the Tomasz Wierzbicki's assumption [23], i.e.,

$$\frac{\sigma_\theta}{\sigma_0} = \frac{M_\theta}{M_0} = \frac{1}{\sqrt{2}}, \quad \frac{\sigma_{ll}}{\sigma_0} = \frac{M_p}{M_0} = \frac{1}{\sqrt{2}}. \tag{6.34}$$

Then, the total bending strain energy rate containing cross-sectional ovalization of a pipe with length l is [24]

$$\dot{W} = \dot{W}_b + \dot{W}_{\text{ovalization}}, \tag{6.35}$$

i.e.,

$$\dot{W} = \frac{M_0 \cdot l\dot{C}}{\sqrt{2}} \cdot \left\{ \left(1 - \frac{2\gamma}{3} - \frac{\gamma^2}{3} \right) + \alpha \cdot \left[\frac{(3+\gamma)}{(1-\gamma)^3} \cdot (1 - 0.95\gamma)^{0.5} \right. \right.$$
$$\left. \left. - \frac{(3-\gamma)}{(1+\gamma)^3} \cdot \left[2 - (1 - 0.95\gamma)^{0.5} \right] \right] \right\}, \tag{6.36}$$

in which

$$\alpha = \frac{\pi t}{32R^2} \frac{\dot{\gamma}}{\dot{C}} \tag{6.37}$$

In Eq. (6.37), α is the adjusted coefficient, which reflects the energetic relation between the pure bending and cross-sectional shape changing of the bending tube, and it is the ratio of the shape-changing rate $\dot{\gamma}$ with respect to the longitudinal bending curvature rate $\dot{C}$.

Equation (6.36) indicates the dependence of energy change rate for a tube bending on the parameter $\dot{C}$ and the adjusted coefficient α. In practice, α will self-adjust to make the total energy required for the pipe bending minimum naturally, i.e.,

$$\frac{\partial \dot{W}}{\partial \gamma} = 0. \tag{6.38}$$

Furthermore, substituting Eq. (6.38) into Eq. (6.36), it obtains

$$\dot{W} = \frac{1}{\sqrt{2}} M_0 \cdot f(\gamma) \cdot l\dot{C}, \tag{6.39}$$

in which, the function $f(\gamma)$ is defined as,

$$
f(\gamma) = \left(1 - \frac{2\gamma}{3} - \frac{\gamma^2}{3}\right) \\
+ \alpha \cdot \left[\frac{(3+\gamma)}{(1-\gamma)^3} \cdot (1 - 0.95\gamma)^{0.5} - \frac{(3-\gamma)}{(1+\gamma)^3} \cdot \left[2 - (1 - 0.95\gamma)^{0.5}\right]\right]. \tag{6.40}
$$

(4) Apparent Bending Moment and Buckling Condition

As the bending tube process is conducted with the action of bending moment M_e, the external moment provides the energy rate $\dot{P}$ on the bending process [24],

$$\dot{P} = M_e \cdot \dot{\varphi} = M_e \cdot l\dot{C}. \tag{6.41}$$

In energy conservation point of view, the consumed strain energy rate within the bending tube should be equal to the external moment rate, i.e.,

$$\dot{P} = \dot{W}. \tag{6.42}$$

Thus, it derives

$$M_e = \frac{1}{\sqrt{2}} M_0 f(\gamma). \tag{6.43}$$

Equation (6.43) indicates that the flattening cross-sectional parameter γ significantly affects the apparent bending moment M_e in the bending process.

The instability of tube bending appears as the curve of M_e versus γ arrives at its peak. It derived the maximum of function $f(\gamma)$ at $\gamma = 0.11$, i.e.,

$$\gamma_c = 0.11. \tag{6.44}$$

Furthermore, it yields the critical longitudinal curvature of the pipe bending at instability status numerically,

$$C_c = \frac{\pi t}{32 R^2} \cdot \int_0^{\gamma_c} \frac{1}{\alpha} \cdot d\gamma = 0.2131 \frac{t}{R^2}. \tag{6.45}$$

Correspondingly, the critical buckling strain of the outer fiber line of the bending tube could be derived [22], which is

$$\varepsilon_c = R(1 - \gamma_c) \cdot C_c = 0.89 R \cdot 0.2131 \frac{t}{R^2} = 0.19 \frac{t}{R}. \tag{6.46}$$

The factor 0.19 in Eq. (6.46) is close to the most experimental results [18, 19, 23].

The comparison of the predictions of Eq. (6.46) with the test data was performed [22], see Fig. 6.3. Meanwhile, the estimations from other approaches are plotted additionally. It should be noted from Fig. 6.3 that the current predictions agree with the available test data as a whole.

Fig. 6.3 Comparison of the available test data with predictions

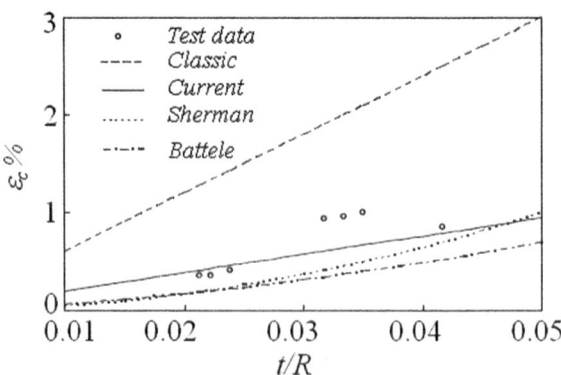

6.5 Effect of Strain Hardening on Plastic Flexural Buckling of Tubes

6.5.1 Effect of Strain Hardening on Critical Buckling Strain of Plastic Bending Tube

Strain hardening or deformation strengthening ability of metallic materials is one of the most important properties of metals, and the most commonly used expression for this ability is the Hollomon formula [25],

$$\sigma = k\varepsilon^n, \tag{6.47}$$

in which, σ and ε present the true stress and true strain, respectively; k is the intensity factor; and n is the hardening exponent of metal.

Okatsu et al. conducted buckling experiments for small-scale tubes of Hollomon-type material, which are with different strain hardening exponent n [25], and the results indicate clearly that lower strain hardening exponent n results in a lower critical buckling strain ε_c, see Fig. 6.4 [25].

As to pipeline steel, it develops from X65 to X100, the ratio of yielding strength to ultimate strength, $R = \sigma_s/\sigma_b$, now becomes an important parameter to characterize its property. It has been found that the higher the ratio of yielding strength to ultimate strength the lower the uniform elongation of metallic material, ε_u, is.

Ji et al. once investigated the tensile properties of X70, X80, and X90 steels. Tensile specimens were cut along the longitudinal tube [26]. The rectangular tensile size of specimens is 50×38.1 mm width, and full wall thickness; the tensile tests were conducted on SHT 4106 machine following ASTM A370. Figure 6.5 shows the tensile properties of some specimens, which reflect clearly that the uniform elongation ε_u for these steels reduces with the increase of the ratio of yield strength

Fig. 6.4 ε_c versus D/t for small-scale tubes [25]

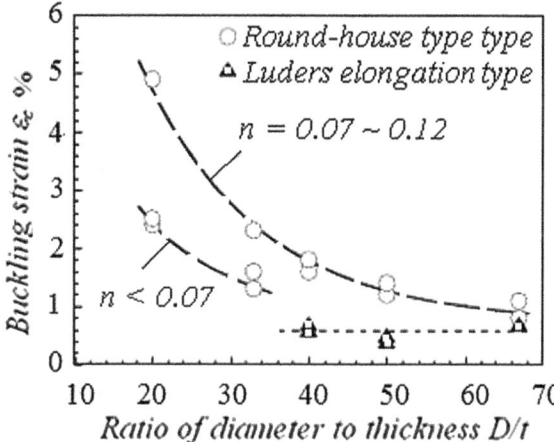

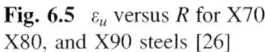

Fig. 6.5 ε_u versus R for X70, X80, and X90 steels [26]

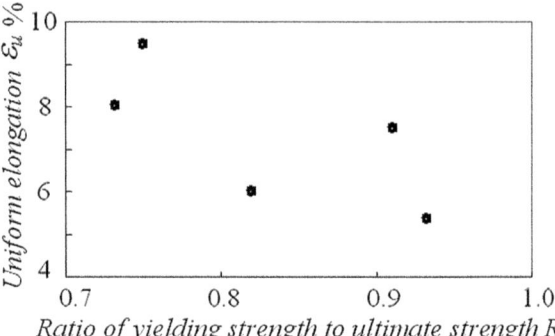

to ultimate strength R. There are other reports which show such phenomenon as well [27].

In addition, the deformability of the steel tube decreases with geometric parameter of tube D/t in hyperbolic form.

Experimental and theoretical analysis indicates that strain hardening exponent is a very important parameter of metallic materials, which controls the strengthening of material during deformation. Theoretically, the value of strain hardening exponent is equal to the maximum value of the uniform strain of material, which reflects the capacity of the material to disperse the deformation outward so as to make deformation uniformly by strain hardening before necking. A correlation which connects the ratio of yield strength to ultimate strength R with the strain hardening exponent n was developed by Hu et al. for Hollomon-type material [28],

$$n = 1 - (\sigma_s/\sigma_b)^{1/2}. \tag{6.48}$$

A lot of experimental data was also gathered by Hu et al. to check the reasonability of Eq. (6.48) [28], and it shows an appropriate agreement, as is shown in Fig. 6.6.

Actually, the strain hardening exponent n reflects the rate of strengthening of Hollomon-type metallic material in deformation process, which reveals the intrinsic correlation between ductility and strength. Higher n corresponds to lower σ_s/σ_b.

The variation of strain hardening exponent n with respect to R was also studied by Jaske for typical Hollomon-type pipeline steels, which is shown in Fig. 6.7 [29].

Figures 6.6 and 6.7 indicate that lower σ_s/σ_b correlates to higher critical buckling strain ε_c obviously.

Fig. 6.6 σ_s/σ_b versus n [28]

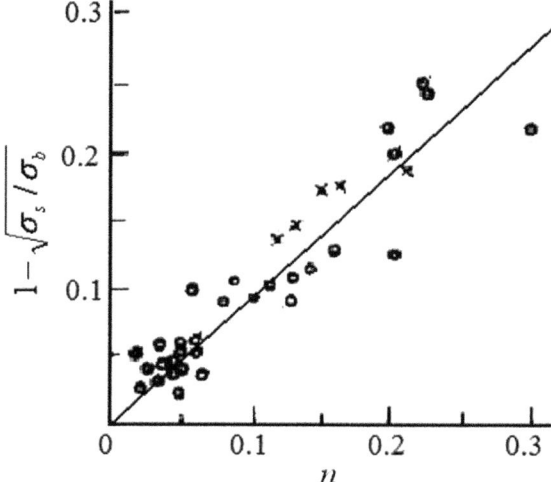

Fig. 6.7 n versus σ_s/σ_b for pipeline steels [29]

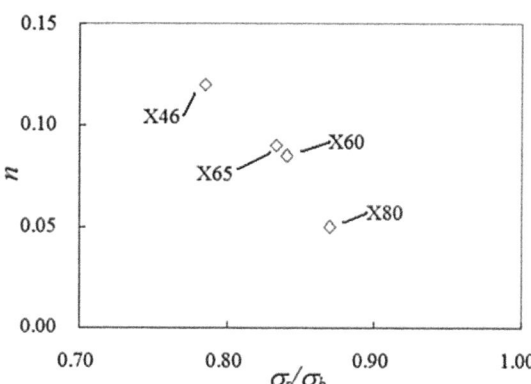

6.5.2   Expression for Critical Buckling Strain Including Strain Hardening Effect for Plastic Bending Tube at Buckling

Yang once investigated the bending moment of a rectangular strain hardening beam analytically [30], and the material exhibits Hollomon-type strain hardening behavior. His result gave an approximate correlation between bending moment of the rectangular strain hardening and $R = \sigma_s/\sigma_b$ [30],

$$
\begin{aligned}
M_P &\approx bh^2\sigma_s \cdot [1+0.904\sigma_b/\sigma_s(1-\sigma_s/\sigma_b)]/4 \\
&= M_0 \cdot [1+0.904\sigma_b/\sigma_s(1-\sigma_s/\sigma_b)],
\end{aligned}
\tag{6.49}
$$

in which $M_0 = bh^2\sigma_s/4$ indicates the bending moment; σ_s is the yielding strength of the beam with ideal plasticity property; h and b are the height and the width of the beam, respectively.

Equation (6.49) reveals clearly that a characteristic factor β can be separated from the expression, which reveals the effect of strain hardening on bending moment of the rectangular beam with Hollomon-type strain hardening behavior especially, i.e.,

$$\beta = [1 + 0.904\sigma_b/\sigma_s(1 - \sigma_s/\sigma_b)]. \tag{6.50}$$

On the other hand, recalling to the three cases of elastically bending buckling condition of pipe, i.e., the 1st one that is the classical (elastic) solution of the bending pipe retaining the cross-section shape perfect round, the Brazier's solution which involves the cross-section shape changing into an elliptical one during of pipe bending, and Li's solution considering the cross section ovalization of pipe due to elastic bending [4, 30, 31], the corresponding critical moment and strain of above models for bending buckling are shown in Table 6.1.

From Table 6.1, it results in a significant consequence that the numerical factors in the classical (elastic) solutions for critical moment and strain are all greater than those in Brazier solution and Li's solution, which is due to the consideration of out-round of the pipe cross-sectional shape in Brazier and Li models. This phenomenon obviously reveals that pipe with better roundness of cross-sectional shape exhibits greater critical buckling moment and strain at the same time during bending, and ratio of the numerical factor ε_c to M_c are all not far from 1.0 in the above three examples.

Meanwhile, the previous section indicates that the function of strain hardening is to ensure deformation uniformly before necking, and it may retain the roundness of cross-sectional shape of pipe during bending. Therefore, as to the Hollomon-type strain hardening material and the bending moment problem, the characteristic factor β separated from bending moment of a rectangular strain hardening beam could be transplanted into the representation for assessing the critical buckling moment and strain of a bending tube to reveal the effect of strain hardening effect due to the similarity of the problem.

Referring that Eq. (6.46) is the estimation of the critical buckling strain of the outer fiber line of a rigid–perfectly plastic bending tube due to cross-sectional ovalization, which is a complete geometric one without considering the action of strain hardening.

Therefore, Eq. (6.46) can be developed to contain the strain hardening effect by the characteristic factor β rationally; therefore, it results in

$$\varepsilon_c = 0.19\frac{t}{r} \cdot \left(1 + \frac{t}{1.78r}\right) \cdot \beta = 0.19\frac{t}{r} \cdot \left(1 + \frac{t}{1.78r}\right) \cdot [1 + 0.904\sigma_b/\sigma_s(1 - \sigma_s/\sigma_b)]$$
$$= 0.19\frac{t}{r} \cdot \left(1 + \frac{t}{1.78r}\right) \cdot (0.096 + 0.904\sigma_b/\sigma_s). \tag{6.51}$$

With the help of Eq. (6.48), the alternate form of Eq. (6.51) is

Table 6.1 Solutions of critical moments and strains for initially circular pipe bending at buckling corresponding three elastic models [4, 31, 32]

Elastic model	Classical (elastic) solution	Brazier solution	Li's solution
M_c	$M_c = 0.577\pi Et^2 r/(1-v^2)^{1/2}$	$M_c = 0.314\pi Et^2 r/(1-v^2)^{1/2}$	$M_c = 0.388\pi Et^2 r/(1-v^2)^{1/2}$
ε_c	$\varepsilon_c = 0.577(t/R)/(1-v^2)^{1/2}$	$\varepsilon_c = 0.366(t/R)/(1-v^2)^{1/2}$	$\varepsilon_c = 0.461(t/R)/(1-v^2)^{1/2}$
Ratio of numerical factor ε_c to M_c	1.000	1.166	1.188
Shape of pipe cross-section during bending	Perfect round	Ellipse	Ovalization

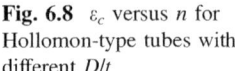

Fig. 6.8 ε_c versus n for Hollomon-type tubes with different D/t

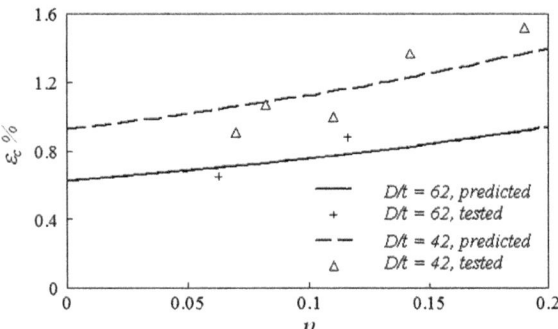

$$\varepsilon_c = 0.19\frac{t}{r} \cdot \left(1 + \frac{t}{1.78r}\right) \cdot \left[0.096 + 0.904/(1-n)^2\right]. \qquad (6.52)$$

Equation (6.52) is the extended expression for critical buckling strain prediction that contains strain hardening effect for plastic bending pipe at buckling.

Ishikawa et al. gathered some data of ε_c versus n for various tubes with varying D/t value of 40–44 and 62 [33]. Figure 6.8 shows the redrawn of the data to reflect the reasonability of Eq. (6.52), in an average value 42 for the D/t = 40–44 is taken. More detailed discussion is in [34]. Figure 6.8 clearly represents the validity of Eq. (6.52).

6.6 Failure of Plastic Flexural Buckling

Above study represents that the expression for critical strain estimation with considering cross-sectional ovalization of bending tube is valid to characterize the plastic bending buckling of a tube. The fundamental assumptions in the derivation include the ellipse cross section of the bending tube and the rigid–perfectly plastic material model.

A characteristic factor $\beta = [1 + 0.904\sigma_b/\sigma_s(1-\sigma_s/\sigma_b)]$ can be separated from the bending moment of the Hollomon-type strain hardening rectangular beam, which reveals the strain hardening effect on bending moment of the rectangular beam as compared to the ideal plastic beam. Furthermore, the effect of strain hardening on the critical buckling strain of a plastic bending tube is contained in the extended expression by analogy method. The material behaves Hollomon-type strain hardening.

References

1. Brazier LG (1927) On the flexure of thin cylindrical shells and other thin sections. Proc Roy Sot Ser A 116:104–114
2. Seide P, Weingarten VI (1961) On the buckling of circular cylindrical shells under pure bending. J Appl Mech ASME 28:112–116

3. Fabian O (1977) Collapse of cylindrical elastic tubes under combined bending, pressure and axial loads. Int J Solids Struct 13:1257–1270
4. Li L (1996) Approximate estimates of dynamic instability of long circular cylindrical shells under pure bending. Int J Press Vessel Pip 61:37–40
5. Jirsa JO, Lee FK, Wilhoit JC, Merwin JE (1970) Ovaling of pipeline under pure bending, OTC 1569. Proceedings Offshore Technology Conference I, pp 573–582
6. Sherman DR (1976) Tests of circular steel tubes in bending. J Struct Div ASCE 102:2181–2195
7. Reddy BD (1979) An experimental study of the plastic buckling of circular cylinders in pure bending. Int J Solids Struct 15:669–683
8. Gellin S (1980) The plastic buckling of long cylindrical shells under pure bending. Int J Solids Struct 16:397–407
9. Bushnell D (1981) Elastic-plastic bending and buckling of pipes and elbows. Comput Struct 13:241–254
10. Calladine CR (1983) Plastic buckling of tubes in pure bending, in Collapse. In: Thompson JMT, Hunt GW (eds) The buckling of structures in theory and practice. Cambridge University Press, Cambridge, pp 111–124
11. Kyriakides S, Shaw PK (1987) Inelastic bending of tubes under cyclic loading. J Press Vessel Tech ASME 109:169–178
12. Wadee MK, Wadee MA, Bassom AP, AndreasAigner A (2006) Longitudinally inhomogeneous deformation patterns in isotropic tubes under pure bending. Proc R Soc A 462:817–838
13. Le Grognec P, Van Le A (2009) Some new analytical results for plastic buckling and initial post-buckling of plates and cylinders under uniform compression. Thin Walled Struct 47:879–889
14. Poonaya S, Teeboonma U, Thinvongpituk C (2009) Plastic collapse analysis of thin-walled circular tubes subjected to bending. Thin Walled Struct 47:637–645
15. Ranzi G, Luongo A (2011) A new approach for thin-walled member analysis in the frame work of GBT. Thin Walled Struct 49:1404–1414
16. Christo MT, Veerappan AR, Shanmugam S (2012) Effect of ovality and variable wall thickness on collapse loads in pipe bends subjected to in-plane bending closing moment. Eng Fract Mech 79:138–148
17. Li HL (2008) Development and application of strain based design and anti-large-strain pipeline steel. Petrol Sci Tech Forum (in Chinese) 27(2):19–25
18. Dorey AB, Murray DW, Cheng JJR (2000) An experimental evaluation of critical buckling strain criteria. In: 2000 international pipeline conference, vol 1. Calgary, Alberta, Canada, 1–5 Oct, pp 71–80
19. Dorey AB, Murray DW, Cheng JJ (2006) Critical buckling strain equations for energy pipelines—a parametric study. Trans ASME 128:248–255
20. Timoshenko SP, Gere JM (2009) Theory of elastic stability, 2nd edn. McGraw-Hill, London, pp 457–485
21. Zheng XL (2004) Mechanical behaviors of engineering materials. Northwestern Polytechnic University Press, pp 8–9
22. Ji LK, Zheng M, Chen HY, Zhao Y, Yu LJ, Hu J, Teng HP (2015) Apparent strain of a pipe at plastic bending buckling state. J Braz Soc Mech Sci Eng 37:1811–1818
23. Yang JL, Reid SR (1997) Approximate estimation of hardening—softening behavior of circular pipes subjected to pure bending. Acta Mech Sin 13(3):227–240
24. Wierzbicki T, Sinmao MV (1997) A simplified model of Brazier effect in plastic bending of cylindrical tubes. Int J Press Vessel Pip 71:19–28
25. Okatsu M, Shinmiya T, Ishikawa N, Kondo J, Endo S (2005) Development of high strength linepipe with excellent deformability. In: 24th international conference on offshore mechanics and arctic engineering: OMAE2005-67149, 3, 63–70. Halkidiki, Greece, pp 12–17
26. Ji LK, Li HL, Wang HT, Zhang JM, Zhao WZ, Chen HY, Li Y, Chi Q (2014) Influence of dual-Phase microstructures on the properties of high strength grade line pipes. J Mater Eng Perform 23(11):3867–3874

27. Toyoda M, Koi M, Hagiwara Y, Seto A (1991) Effects of yield to tensile ratio and uniform elongation on the deformability of welded steel frame structures. Weld Int 5(2):95–101
28. Hu Z, Cao S (1993) Relation between strain—hardening exponent and strength. J Xi'an Jiaotong Univ 27(6):71–76
29. Jaske CE (2002) Development and evaluation of improved model for engineering critical assessment of pipelines. In: Proceedings of international pipeline conference. ASME, New York
30. Yang Z (2006) Plastic limit analysis of beams considering hardening effect. J Huaqiao Univ (Natural Science) 27(3):277–279
31. Yudo H, Yoshikawa T (2015) Buckling phenomenon for straight and curved pipe under pure bending. J Mar Sci Technol 20:94–103
32. Karam GN, Gibson LJ (1995) Elastic buckling of cylindrical shells with elastic cores. I: analysis. Int J Solids Struct 32(8/9):1259–1283
33. Ishikawa N, Okatsu M, Endo S, Kondo J (2006) Design conception and production of high deformability linepipe. In: Proceedings of IPC2006 6th international pipeline conference. Calgary, Alberta, Canada, IPC2006-10240
34. Zheng M, Wei LP, Hu J, Teng HP, Liu J, Wang Y (2019) Critical strain assessment at buckling failure through pipeline bending. Emerg Mater Res 8(5)

Chapter 7
Effect of Defects on Pipe Bending Behavior

Abstract The features of corrosive defects and their effects on conventional properties of pipeline materials are provided first, and then the effect of diffusive defects on pipe bending behavior and the assessments for the limit bending moment of localized corrosive pipeline are comparatively presented in this chapter.

7.1 Defect Types and Simplification

7.1.1 General Introduction

In nowadays, pipeline transportation becomes a very common thing in daily life everywhere; therefore, the safety of pipeline is a serious requirement worldwide. High ductile material and proper prevention from corrosion are the basic demands for pipeline. However, the pipeline transportation has also become one of the most dangerous sources of risk in nowadays due to its high hazard. The economic loss of all equipment failure of petroleum transportation in North America currently costs around $100 billion per year [1]. Liquid medium usually carries harmful acidic material, such as H_2S and CO_2 in petroleum; a corroded pipeline has a relatively lower carrying capacity and safety inevitably, and improper prevention could further leads to accident in the pipeline.

Actually, metal atoms are lost gradually from the pipeline steels during corrosion processes, which leads to a gradationally porous structure in the material along the radial direction of the pipeline due to this corrosion reaction, and the mechanical properties of the pipeline material reduce correspondingly.

A high-resolution magnetic "pig" can be employed to determine the size and position of corrosive defect in a pipeline. Periodic inspections can monitor the growth of defects. Beside, a hydrostatic test is commonly used in guaranty for the safety of pipeline. But actually, this test could not provide a valid guaranty for safety some times due to the "reversal phenomenon." The fact is that a corroded pipeline might fail at a succeeding pressured process though it passes through the hydrostatic test with a higher pressure previously; the occurrence of such phenomenon may be

M. Zheng et al., *Elastoplastic Behavior of Highly Ductile Materials*,
https://doi.org/10.1007/978-981-15-0906-3_7

due to the applied pressure of the hydrostatic test is not far from the failure strength, the pipeline does not fail at that moment, but which results in a significant damage inside the pipeline though it survived, so the pipeline may fail at a succeeding pressured process due to the accumulation of damage continuously to a critical stage. Therefore, the hydrostatic test could not always offer an absolutely safe guarantee of the integrity of the corroded pipeline, even inducing damage instead sometimes.

In general, metal corrosion can be divided into two catalogs, i.e., overall corrosion and localized one. The overall corrosion is also referred to as uniform corrosion. The corrosion reaction is distributed uniformly on the metal surface to certain degrees, and it is difficult to distinguish the cathode and anode of the corroded battery. Generally, the surface is evenly covered with a corrosion product film, which can inhibit corrosion to varying degrees, such as high-temperature oxidation and easy passivation of metal (such as stainless steel, titanium, aluminum, etc.) in the oxidized environment, the passivation film has good protection, even the corrosion process is almost stopped. The overall corrosion is less harmful. While, the localized corrosion is non-uniform, and the corrosion reaction concentrates on the local surface. Localized corrosion can be further divided into galvanic corrosion, small hole corrosion, crevice corrosion, inter-granular corrosion, selective corrosion, stress corrosion cracking, wear corrosion, corrosion fatigue, and hydrogen damage.

Some methods have been developed to estimate the remaining strength of the in servicing pipelines, ASME B31G or CSA Z184, for example. It has been proven by Mok et al. that ASME B31G approach is conservative for long spiral defected pipelines though it satisfies for short corrosion-defected pipelines [2]. The characters of long- and complex-shaped corrosion were studied by Hopkins and Jones [3], and some proposals were presented finally.

As to a corrosion process, it may happen in the radial direction of the pipe step by step from the surface, the corroded layer could be a gradient one from the inner surface of the pipe. So, a "gradient porous material" appears in the radial direction because of the corrosion reaction, which results in the complication of assessing the corroded pipeline. Therefore, it is in need to propose a suitable approach to evaluate the integrity of such a gradient-corroded pipeline.

In morphology or corrosion pattern, corrosive damage can be divided into diffusive type and localized thinning for an actual pipeline.

In the present chapter, the assessment of above corrosion defects on pipeline property is given.

7.1.2 Assessment of Corrosion Damage in Radial Direction of Pipeline

As to the acidic medium, if the inner surface of pipeline is naked during service process, it contacts with harmful materials inevitably, such as CO_2, H_2S, etc., in petroleum. There will be some chemical reactions to consume Fe atoms from the pipeline gradually [4].

The consumption of Fe atoms from the inner surface of pipeline leads to formation of voids (damage) gradually on surface of pipe; simultaneously, there will be a production of hydrogen due to the cathode chemical reactions, which may penetrates into the pipe and concentrates at the boundary of the grains and inclusions-matrix, segregation regions, rolled defects and dislocations, etc.; finally, hydrogen molecules may form. It may even result in a catastrophic consequence such as hydrogen-induced blister or bulge and delimitation due to the accumulation of hydrogen atoms.

In practice, the concentration of such harmful acidic materials, CO_2 and H_2S for example, in the petroleum is not so high generally; so, a diluted solution model is valid to describe such a system. In diluted solution system, there will be no "steric hindrance" or "screening effect," each harmful molecule could perform the chemical reaction with the Fe atom if the essential situation for this chemical reaction occurring is satisfied [4]. On the other hand, the concentration of harmful atoms there changes due to the chemical reaction, which results in the depletion of both Fe atoms and the harmful atoms, i.e. [4],

$$dC = -kCdx \qquad (7.1)$$

in which, C is the content of harmful atoms in the corrosive solution, dC the increment of content of harmful atoms, k is a ratio factor, and dx is the length of material element, respectively. The negative symbol in Eq. (7.1) reflects the decrease of the content induced by the corrosion reaction.

For the corrosion problem in a pipeline, the corrosion occurs from the inner surface and grows in the radial direction. The material element dx in Eq. (7.1) could be set in the radial direction reasonably.

Thus, by integrating Eq. (7.1), the distribution of the harmful atoms in the radial direction of the pipeline is obtained [4],

$$C = C_1 e^{-kx} = C_1 e^{-x/\delta} \qquad (7.2)$$

in which, C_1 is the content of the harmful atoms at the inner surface of the pipe in the solution, x expresses the radial distance from the inner surface of the pipe, $\delta = 1/k$ is the parameter which reflects the penetration depth of corrosive damage.

7.2 Treatment of Diffusive Defects

The corrosive chemical reaction induces the depletion of Fe atoms directly from the pipeline, and void-type damage is caused corresponding due to the loss of Fe atoms; the void-type damage is proportional to the local content of harmful atoms in the solution inevitably, therefore the distribution of such damage in the radial direction of the pipeline can be approximated as [4],

$$D = D_1 e^{-kx} = D_1 e^{-x/\delta} \tag{7.3}$$

D_1 is the damage variable at the inner surface of the pipeline, which is in void type.

In 1991, Zheng et al. investigated the influence of voids on elastic modulus of materials [5], and found that the effect of voids on elastic modulus of materials can be expressed by following formula [5]

$$E = E_0(1 - 2.625 \phi) = E_0(1 - D) \tag{7.4}$$

in which E_0 is the elastic modulus of the original material without void, ϕ is the void fraction, and $D = 2.625\phi$ expresses the damage variable [5].

Substituting Eq. (7.3) into Eq. (7.4), it yields the distribution of effective elastic modulus in the radial direction,

$$E = E_0(1 - D_1 e^{-x/\delta}) = (E_1 - E_0)e^{-x/\delta} + E_0 \tag{7.5}$$

where $E_1 = E_0(1 - D_1)$ is the effective elastic modulus of the damaged material at the inner surface of the pipeline.

In addition, according to Zheng's study about the effect of voids on yielding strength of materials [6], the effective yielding strength of the material with randomly distributed voids is,

$$\sigma_s = \sigma_{s0}[(1 - \phi)(1 - 2.625\phi)/(1 + 6.25\phi \, \sigma_m^2/\sigma_s^2)]^{1/2}, \tag{7.6}$$

For uniaxial tensile loading [6], the stress triaxiality $\sigma_m/\sigma_s = 1/3$, Eq. (7.6) becomes

$$\sigma_s = \sigma_{s0}[(1 - \phi)(1 - 2.625\phi)/(1 + 6.25\phi/9)]^{1/2}. \tag{7.7}$$

Besides, according to Zheng's study of the effect of voids on fracture toughness of materials [7], fracture toughness of porous metal is

$$K_{IC} = \left[(1 - 2.625\phi)(1 - \phi^{2/3})\right]^{1/2} K_{IC0}, \tag{7.8}$$

in which K_{IC0} is the toughness of metal free of voids.

7.3   Effect of Diffusive Defects on Pipe Bending Behavior

For a plate with the thickness t, if the penetration depth of corrosive damage in the plate is δ, the effective elastic modulus along the longitudinal direction in the damaged region from Eq. (7.5) is,

$$E' = \frac{1}{\delta} \int_0^\delta E \cdot \mathrm{d}x = \frac{1}{\delta} \int_0^\delta E_0 \left(1 - D_1 \mathrm{e}^{-x/\delta}\right) \mathrm{d}x = E_0 \left[1 - D_1\left(1 - \mathrm{e}^{-1}\right)\right]. \quad (7.9)$$

While, the effective elastic modulus of the whole plate along the longitudinal direction is,

$$E = E' \cdot \delta/t + E_0 \cdot (t - \delta)/t = E_0 - E_0 \left[D_1\left(1 - \mathrm{e}^{-1}\right)\right] \cdot \delta/t. \quad (7.10)$$

Similarly, the effective yielding strength along the longitudinal direction in the damaged region from Eq. (7.7) is,

$$\sigma'_s = \frac{1}{\delta} \int_0^\delta \sigma_s \mathrm{d}x = \frac{1}{\delta} \int_0^\delta \sigma_{s0} [(1 - \phi)(1 - 2.625\phi)/(1 + 6.25\phi/9)]^{1/2} \cdot \mathrm{d}x \quad (7.11)$$

The direct integration in Eq. (7.11) is a little bit difficult, thus a simplified approximation is needed. An approximated expression of $[(1 - \phi)(1 - 2.625\phi)/(1 + 6.25\phi/9)]^{1/2}$ is $(1 - 2.625\phi)$, as is shown in Fig. 7.1, which is a conservative approximation.

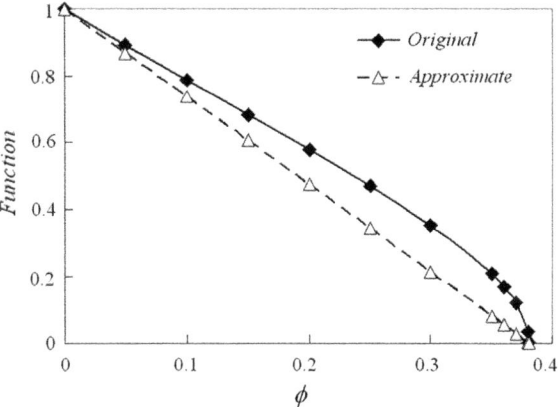

Fig. 7.1 Comparison of approximated expression $(1 - 2.625\phi)$ with original expression $[(1 - \phi)(1 - 2.625\phi)/(1 + 6.25\phi/9)]^{1/2}$

Thus, a conservative estimation for Eq. (7.11) is

$$
\begin{aligned}
\sigma'_s &= \frac{1}{\delta} \int_0^\delta \sigma_s \mathrm{d}x \approx \frac{1}{\delta} \int_0^\delta [\sigma_{s0}(1 - 2.625\phi)] \cdot \mathrm{d}x \\
&= \frac{1}{\delta} \int_0^\delta [\sigma_{s0}(1 - D)] \cdot \mathrm{d}x = \frac{1}{\delta} \int_0^\delta \left[\sigma_{s0}\left(1 - D_1 e^{-x/\delta}\right)\right] \cdot \mathrm{d}x \\
&= \sigma_{s0}[1 - D_1(1 - e^{-1})]
\end{aligned}
\tag{7.12}
$$

Simultaneously, the effective yielding strength of the whole plate along the longitudinal direction is,

$$
\sigma_s = \sigma'_s \cdot \delta/t + \sigma_{s0} \cdot (t - \delta)/t = \sigma_{s0}\left\{1 - \left[D_1(1 - e^{-1})\right] \cdot \delta/t\right\}.
\tag{7.13}
$$

Furthermore, in elastic case, the elastic bending moment of such corrosively damaged pipe can be written as,

$$
M = \pi E R^3 t/\rho = \pi E_0\left\{1 - \left[D_1(1 - e^{-1})\right] \cdot \delta/t\right\} R^3 t/\rho,
\tag{7.14}
$$

in which R, t, and ρ express the radius, the thickness and the curvature radius of the bending pipe.

In plastic case, the plastic bending moment of such corrosively damaged pipe is,

$$
M = 4\sigma_s \cdot R^2 t = 4\sigma_{s0}\left\{1 - \left[D_1(1 - e^{-1})\right] \cdot \delta/t\right\} \cdot R^2 t,
\tag{7.15}
$$

Equations (7.14) and (7.15) indicates that the bending moments of a corrosively gradient damage pipe in both elastic and plastic cases decrease inevitably.

7.4 Estimation for the Limit Bending Moment of Localized Corrosive Pipeline

It is a very common to form a volume defect and local wall thinning in pressure vessels and pipes, which can be caused by corrosion and mechanical damage. Current research on local wall thinning mainly focuses on pipes under internal pressure [8].

Investigations have developed analytical methods for the assessment of the remaining strength of pipes with local wall thinning [9–15]. ASME/B31.G is the most widely admitted approach for this assessment [9]. However, it is only satisfied as an assessment for corroded pipes under internal pressure condition; while evaluation of the effect of local thinning on bending behavior of the corroded pipe remains unsolved.

Bending and axial loads inevitably exist in engineering practice, especially for oil and gas transmission. Therefore, appropriate evaluation of the effect of local thinning on bending behavior of the corroded pipe is needed.

In this section, the modified expression for assessing the limit moment of a pipe with local wall thinning is presented, which takes the effects of longitudinal defect length on the limit bending moment into account; the comparison of finite element analysis and the net-section collapse (NSC) criterion verified the validity of the predictions from the modified expression [8, 9, 16].

7.4.1 Discrepancy of NSC and FEM Analysis for Assessing Limit Bending Moment

Kanninen et al. once proposed a net-section collapse failure (NSC) criterion to deal with pipe leak and break loads as indicating critical net-stresses at crack initiation and maximum load [17]. Furthermore, the failure analysis of a locally thinned pipe under bending was conducted by using NSC criterion. The additional condition for this treatment is an assumption that distribution of bending stress is uniform in the component (Fig. 7.2). It further supposes that collapse appears as the net-section stress of the structure arrives at the flow stress of the pipe material, σ_0.

In engineering practice, the thickness t of the pipe is much smaller than the radius R of the pipe, i.e., $t/R \ll 1$. Therefore, it is reasonable to ignore the wall thickness effect on the assessment. Under this assumption, the radius of the pipe is characterized by one parameter, R only. Furthermore for the sake of simplicity, the local wall thinning is assumed at the projection center on the diameter plane, which is consistent with the maximum bending moment plane. The limit load of the local wall thinned pipe could be obtained from equilibrium of bending moments and axial force. Under condition of $\theta + \beta < \pi$, the limit moment can be written as readily [16, 17],

$$\beta = \frac{\pi}{2}\left(1 - \frac{(1-\eta)\theta}{\pi} - \frac{p}{2\pi R_m t \sigma_0}\right), \tag{7.16}$$

$$M \approx 2\sigma_0 R_m^2 t[2\sin\beta - (1-\eta)\sin\theta]. \tag{7.17}$$

Fig. 7.2 Sketch of the part of the local thinned pipe

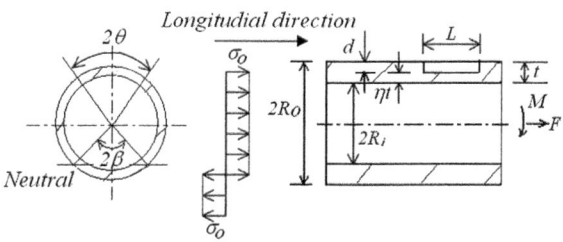

While in case of $\theta + \beta < \pi$ [16, 17],

$$\beta = \frac{\pi}{2\eta} \left(2\eta - \frac{(1-\eta)\theta}{\pi} - 1 - \frac{p}{2\pi R_m t \sigma_0} \right), \tag{7.18}$$

$$M \approx 2\sigma_0 R_m^2 t [2\eta \sin \beta - (1 - \eta) \sin \theta]. \tag{7.19}$$

in which θ expresses the semi-angle of the local wall thinned part, β is the semi-angle at the neutral axis (Fig. 7.2), R_m is the mean radius of the pipe, and $\eta = (t-d)/t$; the flow stress, the axial load, the limit moment, and the limit moment of a plain pipe are expressed as σ_0, P, M, and M_0, respectively.

If there is no axial load, i.e. $P = 0$, Eqs. (7.16)–(7.19) can be simplified as, in the case of $\theta + \beta < \pi$,

$$\beta = \frac{\pi}{2} \left(1 - \frac{(1-\eta)\theta}{\pi} \right), \tag{7.20}$$

$$M \approx 4\sigma_0 R_m^2 t \left[\sin \beta - \frac{(1-\eta)}{2} \sin \theta \right] = M_0 \cdot \left[\sin \beta - \frac{(1-\eta)}{2} \sin \theta \right]. \tag{7.21}$$

While in case of $\theta + \beta < \pi$,

$$\beta = \frac{\pi}{2\eta} \left(2\eta - \frac{(1-\eta)\theta}{\pi} - 1 \right), \tag{7.22}$$

$$M \approx 2\sigma_0 R_m^2 t \left[\eta \sin \beta - \frac{(1-\eta)}{2} \sin \theta \right] = M_0 \cdot \left[\eta \sin \beta - \frac{(1-\eta)}{2} \sin \theta \right]. \tag{7.23}$$

Han et al. employed 3D elastic–plastic finite element analysis (FEM) to show the effect of the longitudinal size of the defect on the limit bending moment [11].

In the proposed 3D models [16], a bending load was applied by four-point bending. A quadrant of the pipe was taken and the appropriate symmetry boundary conditions were applied. The local wall thinning was assumed and located in the center of pipe; the radius ratio of outer to inner R_0/R_i was 1.2.

The EMRC/NISA program with 20-node iso-parametric elements is employed in the finite element analysis [16].

The limit moment was defined by the convergent solution of load for a perfectly plastic material in the nonlinear procedure. The material properties including Young's modulus, flow stress, and Poisson's ratio in the analysis were as follows: $E = 210$ GPa, $\sigma_f = 200$ MPa, and $v = 0.3$, respectively [16].

The finite element analysis results indicated a significant dependence of limit bending moment on the relative longitudinal length of the defect; however, the net-section collapse criterion (NSC) proposed by Kanninen showed that the limit bending moment is independent on the longitudinal relative length of the defect.

Above fact indicates the incompleteness of the NSC in describing the effect of defect size. Therefore, an appropriate modification to the NSC is needed.

7.4.2   Effective Thickness Model for Predicting the Limit Bending Moment of Locally Corroded Pipe

Zheng et al. once proposed an effective thickness model to predict the limit bending moment of locally corroded pipe [8].

In the effective thickness model, the wall thickness t is corrected as the effective thickness t_{corr}, which was formulated empirically by means of burst failure results [8, 9]

$$t_{corr} = \frac{t - d}{1 - \dfrac{d}{t\left\{1 + 0.8\left[L/(Dt)^{0.5}\right]^2\right\}^{0.5}}}. \qquad (7.24)$$

In Hauch and Bai's research for the initial plastic yielding of a non-uniform cylindrical shell in Haagsma and Shaap's formula, the effective thickness t_{corr} was also used to characterize the effect of defect length [18, 19], and reasonable results were obtained.

As a result, the thickness in the defected area of the shell in Kanninen's net-section collapse failure (NSC) criterion was modified by the 'effective thickness,' t_{corr} of Eq. (7.24). Thus, a modified η was derived as [8]

$$\eta_{corr} = \frac{1 - d/t}{1 - \dfrac{d}{t\left\{1 + 0.8\left[L/(Dt)^{0.5}\right]^2\right\}^{0.5}}} = \frac{1 - d/t}{1 - \dfrac{d}{t\left\{1 + 0.4\left[L/(Rt)^{0.5}\right]^2\right\}^{0.5}}}, \qquad (7.25)$$

where $D = 2R$.

Furthermore, under condition of $\theta + \beta < \pi$, the modified expressions for limit bending moment was derived [8],

$$\beta = \frac{\pi}{2}\left(1 - \frac{(1 - \eta_{corr})\theta}{\pi} - \frac{p}{2\pi R_m t \sigma_0}\right), \qquad (7.26)$$

$$M \approx 2\sigma_0 R_m^2 t[2\sin\beta - (1 - \eta_{corr})\sin\theta]. \qquad (7.27)$$

While in case of $\theta + \beta > \pi$ [8],

$$\beta = \frac{\pi}{2\eta_{corr}}\left(2\eta_{corr} - \frac{(1 - \eta_{corr})\theta}{\pi} - 1 - \frac{p}{2\pi R_m t \sigma_0}\right), \qquad (7.28)$$

$$M \approx 2\sigma_0 R_m^2 t[2\eta_{corr}\sin\beta - (1 - \eta_{corr})\sin\theta]. \qquad (7.29)$$

Again, if there is no axial load, i.e. $P = 0$; Eqs. (7.26)–(7.29) were simplified as, in the case of $\theta + \beta < \pi$ [8],

$$\beta = \frac{\pi}{2}\left(1 - \frac{(1 - \eta_{corr})\theta}{\pi}\right), \tag{7.30}$$

$$M \approx 2\sigma_0 R_m^2 t[2\sin\beta - (1 - \eta_{corr})\sin\theta] = M_0\left[\sin\beta - \frac{(1 - \eta_{corr})}{2}\sin\theta\right]. \tag{7.31}$$

While in case of $\theta + \beta > \pi$ [8],

$$\beta = \frac{\pi}{2\eta_{corr}}\left(2\eta_{corr} - \frac{(1 - \eta_{corr})\theta}{\pi} - 1\right), \tag{7.32}$$

$$M \approx 4\sigma_0 R_m^2 t\left[\eta_{corr}\sin\beta - \frac{(1 - \eta_{corr})}{2}\sin\theta\right] = M_0\left[\eta_{corr}\sin\beta - \frac{(1 - \eta_{corr})}{2}\sin\theta\right]. \tag{7.33}$$

The comparison results showed the validity of Eqs. (7.30)–(7.33) [8].

7.5 Summary

Diffusive defects (damage) due to corrosion result in the decrease of elastic modulus, the plastic yielding strength, and the bending moments of a damaged pipe in both elastic and plastic cases inevitably.

The longitudinal thinned length has a significant influence on the limit bending moment of a pipe under bending and tension loads, the effect becomes remarkable and tends to a steady value as the relative defect length increases over $L/(Rt)^{0.5} > 1.5$. The incompleteness of Kanninen's net-section collapse failure (NSC) criterion is that the limit bending moment is independent on the relative longitudinal length of the defect. The modified NSC expressions could reflect the influence of the relative defect length on the limit bending moment and give reasonable results.

References

1. Timmins PF (1997) Solutions to hydrogen attack in steel. ASM International Materials Park, USA, pp 1–10
2. Mok DR, Pick RJ, Glover AG (1990) Behavior of line pipe with long external corrosion. Mater Performance 29:75–79

3. Ahammed M (1998) Probability estimation of remaining life of a pipeline in the presence of active corrosion defect. Int J Press Pip 75:321–329
4. Zhao XW, Luo JH, Zheng M, Lu MX, Li HL (2002) A damage model for assessing pipeline safety in corrosion environments. Met Mater Int 8(5):479–485
5. Zheng M, Zheng X (1991) Expression for predicting the elasticity modulus of materials reinforced by 2nd phase grains. Metall Trans A 22:507–511
6. Zheng M, Luo ZJ, Zheng X (1992) The yielding behavior of materials with random voids. Chinese Sci Bull 37:512–516
7. Zheng M, Luo ZJ, Zheng X (1994) Intensity and toughness parameters of porous materials. Chinese Sci Bull 39:810–814
8. Zheng M, Luo JH, Zhao XW, Zhou G, Li HL (2004) Modified expression for estimating the limit bending moment of local corroded pipeline. Int J Press Vessels Pip 81:725–729
9. ANI/ASME B31.G (1991) Manual for determining the remaining strength of corroded pipeline
10. CAN/CSA-Z184-M86 (1986) Gas pipeline system, Canadian Standards Association
11. Folias ES (1965) An axial crack in a pressurized cylindrical shell. Int J Fract 1(2):104–113
12. Anon (1999) DNV-RP-F101 corroded pipelines, Det Noritas
13. Klever FJ (1992) Burst strength of corroded pipe: flow stress revised. In: Offshore technology conference, Houston, Texas, May 4–7 1992
14. Klever FJ, Stewart G, van der Valik CAC (1995) New developments in burst strength predictions for locally corroded pipeline, 1995 OMAE, V. Pipeline Technology. ASME
15. Fu B, Kirkwood MG (1995) Predicting failure pressure of internally corroded linepipe using the finite element methods, 1995 OMAE, V. Pipeline Technology. ASME
16. Han LH, He SY, Wang YP, Liu CD (1999) Limit moment of local wall thinning in pipe under bending. Int J Press Vessels Pip 76:539–542
17. Kanninen MF, Broek D, Hahn GT, Marschall CW, Rybicki EF, Wilkowski GM (1978) Toward an elastic fracture mechanics predictive capability for reactor piping. Nucl Eng Des 48:117–134
18. Hauch S, Bai Y (1998) Use of finite element methods for the determination of local buckling strength. In: Proceedings of the 1998 international conference on offshore mechanics and Arctic Engineering, Lisbon, Portugal, 5–9 July 1998
19. Haagsma SC, Schaap D (1981) Collapse resistance of submarine line studies. Oil Gas J 2:86–95

Chapter 8
Thermal Stress Problems

Abstract The general description of thermal stress problem is given first, then the thermal stress in granular reinforced composite is presented, and the double embedding model-based analysis is addressed especially; SiC grain-reinforced Al matrix composite is taken as example to show the details of thermal stress distribution and evolution during temperature change particularly.

8.1 Conditions of Thermal Stress Formation

In general, material expands or contracts as temperature changes, and in a relatively wide range of temperatures, this expansion or contraction is proportional to the variation of temperature.

This proportionality can be expressed by a coefficient, called *linear thermal expansion coefficient* α, which is defined as the relative change in length per temperature change, i.e., $1°$.

If the expansion or contraction of material is in the uniform temperature field freely, all the fibers inside the material is free without any constraint, no stress is caused by the change in temperature.

However, when the temperature change is not uniform in a homogeneous body, different material element tends to expand by different amount according to its local temperature change, and the whole body remains continuous which conflicts with the requirement that each element expands to itself amount. Thus, the various elements exert upon each other a restraining action, which results in continuous unique displacement at every point. The strains produced in this system cancel out all or part of the free thermal expansions at every point so as to ensure continuity of displacement. Therefore, this strain system is accompanied by a corresponding system of self-equilibrating stresses certainly. These stresses are known as *thermal stress*.

A similar stress system may be induced in a structure consisting of dissimilar materials even when the temperature change throughout the structure is uniform.

© Springer Nature Singapore Pte Ltd. 2019
M. Zheng et al., *Elastoplastic Behavior of Highly Ductile Materials*,
https://doi.org/10.1007/978-981-15-0906-3_8

Also, if the temperature change in a homogeneous body is uniform and external restraints limit the amount of expansion or contraction, the stresses produced in the body are termed also *thermal stress*.

8.2 Balanced Differential Equations

The idea of *thermal stress* can be illustrated by the following simple examples. Two parallel plates of different materials with the same length L are fixed together along one edge at temperature T_1, this is the initial stress-free state, see Fig. 8.1. Then at a different temperature T_2, there is a temperature rise $\Delta T = T_2 - T_1$. Each plate attempts to expand its amount according to itself nature, i.e., plate 1 tries to expand to an amount of $\delta_1 = \alpha_1 L \Delta T$, while plate 2 attempts to expand to an amount of $\delta_2 = \alpha_2 L \Delta T$. However, the two plates are fixed together, they could only expand to a common amount of δ so as to ensure the continuity of displacement. Thus, there is inconsistent between the "attempting expansion amount" and the "actual expansion amount," which induces the thermal stress inside the plates.

The conditions of equilibrium and compatibility of strain are given, respectively, by

$$\sigma_1 A_1 + \sigma_2 A_2 = 0 \tag{8.1}$$

$$E_1(\delta_1 - \delta)/L = \sigma_1, \ E_2(\delta_2 - \delta)/L = \sigma_2 \tag{8.2}$$

in which, σ_1 and σ_2 are the corresponding tensile thermal stresses in the two plates; A_1 and A_2 express the cross-sectional areas of the plates, respectively; E_1 and E_2 indicate the elastic modulus of the two plates; The term $\alpha L \Delta T$ refers to the free thermal expansion, while the terms $\alpha \Delta T$ refer to the corresponding strain.

The solution of the above equations yields the following results for σ and δ,

$$\delta = (E_1 A_1 \delta_1 + E_2 A_2 \delta_2)/(E_1 A_1 + E_2 A_2), \tag{8.3}$$

$$\sigma_1 = E_1 E_2 A_2(\delta_1 - \delta_2)/[L(E_1 A_1 + E_2 A_2)] = E_1 E_2 A_2(\alpha_1 - \alpha_2)\Delta T/(E_1 A_1 + E_2 A_2), \tag{8.4}$$

$$\sigma_2 = E_1 E_2 A_1(\delta_2 - \delta_1)/[L(E_1 A_1 + E_2 A_2)] = E_1 E_2 A_2(\alpha_2 - \alpha_1)\Delta T/(E_1 A_1 + E_2 A_2). \tag{8.5}$$

Fig. 8.1 Simple two-plate fixed structure

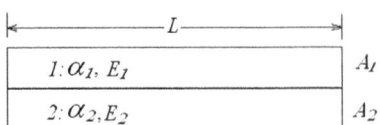

From Eqs. (8.4) and (8.5), for a composite structure consisting of dissimilar materials and with a uniform temperature rise, it can be seen that the self-equilibrating thermal stresses depend not only upon the temperature differences in the structure, but also upon the geometry (A) of the constituting structural components and materials properties.

On the other hand, if the unrestrained structure is homogeneous, thermal stresses could result from non-uniform temperature change, besides the occurring of thermal stress in a non-homogeneous structure with uniform temperature change.

In general, the thermal stress is influenced by the temperature level due to temperature-dependence of material properties, and the material might be a non-homogeneous, elastic, or elastic–plastic medium. Usually, it is reasonable to assume that the material is elastic or elastic–plastic status so as to perform thermal stress analysis for many cases.

In the above simple discussion of thermal stresses assessment, it assumed that the temperature distribution was known. However, in many cases the distribution of temperature should also be conducted simultaneously in proper manner by means of *mechanics* and *thermodynamics*.

In general, there are six stress compatibility equations as well. Two of these six stress compatibility equations are presented here in Cartesian coordinate system, the remaining four can be written down with appropriate change of the variable,

$$(1+v)\nabla^2\sigma_{xx} + \frac{\partial^2\Sigma}{\partial x^2} + E\left[\frac{1+v}{1-v}\nabla^2(\alpha T) + \frac{\partial^2(\alpha T)}{\partial x^2}\right]$$
$$+ v\left(\frac{1+v}{1-v}\right)\left(\frac{\partial X}{\partial x} + \frac{\partial Y}{\partial y} + \frac{\partial Z}{\partial z}\right) + 2(1+v)\frac{\partial X}{\partial x} = 0 \tag{8.6}$$

$$(1+v)\nabla^2\sigma_{xz} + \frac{\partial^2\Sigma}{\partial x\partial z} + E\frac{\partial^2(\alpha T)}{\partial x\partial z}$$
$$+ (1+v)\left(\frac{\partial X}{\partial z} + \frac{\partial Z}{\partial x}\right) = 0 \tag{8.7}$$

in which, Σ is defined by $\Sigma = \sigma_{xx} + \sigma_{yy} + \sigma_{zz}$; X, Y, and Z are volume force density in x, y and z directions, respectively; v is Poisson's ratio.

In general, the thermoelastic problem is reduced to solve the three equilibrium equations in terms of the displacements together with the appropriate boundary conditions. The displacement functions are given by u, v, and w throughout the body. A typical equation for this formulation is shown here; the other two can be written down with appropriate change of variable,

$$(\lambda+\mu)\frac{\partial\bar{e}}{\partial x} + \mu\nabla^2 u - \beta\frac{\partial T}{\partial x} + X = 0 \tag{8.8}$$

$$\bar{e} = \frac{\partial u}{\partial x} + \frac{\partial v}{\partial y} + \frac{\partial w}{\partial z} \tag{8.9}$$

where λ and μ, the Lamé elastic constants, which are defined by

$$\lambda = vE/(1+v)(1-2v); \; \mu = E/2(l+v) = G; \; \beta = E\alpha/(1-2v) \qquad (8.10)$$

8.3   Examples of Thermal Stress Problems

Example 1 For a uniform-thickness free-thin plate (Fig. 8.2), $l \gg c$, and the initial temperature along the thickness of the plate is uniform. Furthermore, there is uneven variation of temperature along the height, that is, $T = T(y)$. Solve the thermal stress in this plate.

Solution The sheet belongs to a free boundary problem, i.e., there is no external constraint. Since it has an uneven temperature variation along y-direction, the fibers in the layers will have different changes in length in x-direction. In order to meet the deformation coordination conditions, the deformation of the fibers of each layer will be restrained by the nearby fibers, so it will generate thermal stress in the plate. Since the plate is with $l \gg c$, and the temperature is independent of x, it can be considered as a one-dimensional problem, and there is only the stress σ_{xx} in x-direction in the plate.

It is assumed that the ends of the thin plate are fixed before the temperature change first, and the longitudinal fibers are not deformed after the temperature change. From the Eq. (8.4), the stress that will appear in the plate is

$$\sigma'_x = -E\alpha T(y) \qquad (8.11)$$

While the restraining reaction force appears at the two fixed ends. The resultant force of the thin plate in the x-direction and the moment about the z-axis can be obtained by the Eq. (8.11), that is,

$$N_x = \int_{-c}^{c} \sigma'_x dy = - \int_{-c}^{c} E\alpha T(y) dy \qquad (8.12)$$

Fig. 8.2 Uniform-thickness free-thin plate

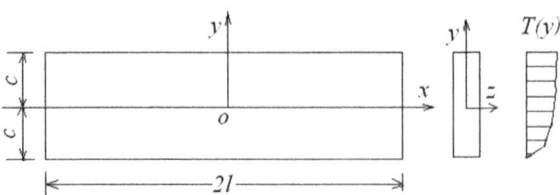

$$M_z = \int_{-c}^{c} \sigma'_x y\,dy = -\int_{-c}^{c} E\alpha T(y) y\,dy \tag{8.13}$$

In order to relieve the displacement constraint at both ends of the thin plate so as to meet the conditions of the free boundary, the axial force and moment opposite to the symbols of Eqs. (8.12) and (8.13) must be applied at the end, and the stress in the plate is superposition of the Eq. (8.11) and the stresses produced from Eqs. (8.12) and (8.13).

$$\sigma_x = \sigma'_x + \sigma''_x + \sigma'''_x = -E\alpha T(y) + \frac{E\alpha}{2c}\int_{-c}^{c} T(y)\,dy + \frac{3y}{2c^3}E\alpha\int_{-c}^{c} yT(y)\,dy \tag{8.14}$$

If the temperature change $T(y)$ is a linear function, that is, $T = a + by$, then substituting it into the Eq. (8.14), it yields $\sigma_x = 0$, that is, there is no thermal stress in the plate. It can be proved (omitted here): if there is no displacement constraint and boundary force on the boundary of the object, and there is no force density within the body, there will be no thermal stress generated in the object as the temperature change in the object is a linear function of the coordinate. Furthermore, according to the superposition principle, in an object with a free boundary, if there is no force density within the body, "superimposing a linear distribution temperature field" on the original temperature field, the stress distribution of the object will not be changed, but the deformation of the object will change.

Example 2 Inside the large sphere of radius b, there is a small sphere with radius a, which is placed at the center, and $a \ll b$, the temperature of the small sphere is T, the thermal expansion coefficient of the material is α, the elasticity modulus is E. Find the thermal stress.

Solution First, assuming that the small sphere is completely unconstrained, its radial expansion is

$$\Delta a' = \alpha T a \tag{8.15}$$

In fact, the expansion of the small sphere will be bound by the big sphere because it is a part of the big sphere, and it is far away from the outer boundary. Considering the spherical symmetry, a uniform pressure p will be applied to the surface of the small sphere. The resulting radial strain is $-p(1 - 2v)$, and the actual amount of change in radius is

$$\Delta a'' = -(1 - 2v)pa/E \tag{8.16}$$

Therefore, the amount of change in the radius of the small sphere should be

$$\Delta a = \Delta a' + \Delta a'' = \alpha T a - (1 - 2v)pa/E \tag{8.17}$$

The effect of the small sphere on the big sphere is transparent. The inner surface of the big sphere is subjected to a uniform pressure p, and the stress distribution can be determined by the following formula,

$$\sigma_r = \frac{a^3(r^3 - b^3)}{r^3(b^3 - a^3)}p, \ \sigma_\theta = \sigma_\varphi = \frac{a^3(2r^3 + b^3)}{2r^3(b^3 - a^3)}p \tag{8.18}$$

Considering $a \ll b$, as the stress distribution near $r = a$ is mainly studied, the above formula can be approximated as

$$\sigma_r \approx -\frac{a^3}{r^3}p, \ \sigma_\theta = \sigma_\varphi \approx \frac{a^3}{2r^3}p \tag{8.19}$$

At the inner surface of the large sphere ($r = a$), the stress component is

$$\sigma_r = -p, \ \sigma_\theta = \sigma_\varphi = p/2 \tag{8.20}$$

If u is the radial displacement of the large sphere, the geometric relationship is

$$\varepsilon_\theta = \varepsilon_\varphi = u/r \tag{8.21}$$

And there is

$$\varepsilon_\theta = \varepsilon_\varphi = [(1 - v)\sigma_\theta - v\sigma_r]/E \tag{8.22}$$

The radial displacement incremental at $r = a$ is

$$\Delta a = (a\varepsilon_\theta)|_{r=a} = (a/E) \cdot [(1 - v)\sigma_\theta - v\sigma_r]|_{r=a} \tag{8.23}$$

It obtains from Eqs. (8.17) and (8.23)

$$p = 2E\alpha T/[3(1 - v)] \tag{8.24}$$

Substituting Eq. (8.24) into Eq. (8.19), the stress component in the outer portion of the small sphere (i.e., the large sphere) is obtained,

$$\sigma_r = -\frac{2\alpha E T a^3}{3(1 - v)r^3}, \ \sigma_\theta = \sigma_\varphi = \frac{\alpha E T a^3}{3(1 - v)r^3}. \tag{8.25}$$

Example 3 As shown in Fig. 8.3, a rectangular thin plate with a peripherally fixed shape has a thermal expansion coefficient of α; the elastic modulus is E; and the

Fig. 8.3 Fixed rectangular
thin sheet

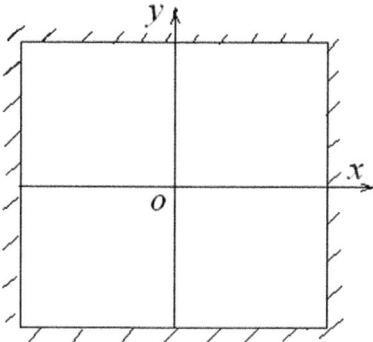

Poisson's ratio is v. When the temperature of the thin plate is raised T, find the
thermal stress inside the plate.

Solution As to a flat plate in a uniform temperature field, if there is no external
constraint, the thermal expansion induced by temperature rise is free, and thus no
thermal stress generates in the plate. In fact, the problem now is that the plate is
fixed, the thermal expansion in x- and y-directions is limited. So there will generate
thermal stress in the plate and it is in a bidirectional stress state, which can be
represented by σ_x and σ_y. Since the plate is fixed, each micro-element does not
deform freely in x- and y-directions, that is,

$$\varepsilon_x = \varepsilon_x^e + \varepsilon_x^T = 0, \ \varepsilon_y = \varepsilon_y^e + \varepsilon_y^T = 0 \tag{8.26}$$

where ε_x^e and ε_y^e are elastic strains caused by thermal stress; ε_x^T and ε_y^T are strains
caused by temperature change. Considering the thin plate is free in z-direction, let
$\sigma_z = 0$, and remind the conditions given by Eq. (8.26), then there are

$$\varepsilon_x = \left(\sigma_x - v\sigma_y\right)/E + \alpha T = 0, \ \varepsilon_y = \left(\sigma_y - v\sigma_x\right)/E + \alpha T = 0. \tag{8.27}$$

The solutions of Eq. (8.27) are,

$$\sigma_x = \sigma_y = -E\alpha T/(1 - v) \tag{8.28}$$

In Eq. (8.28), there is a negative sign which indicates that the thermal stress in
the plate is compressive when the temperature rises $T > 0$. Comparing Eq. (8.27)
with fixed one-dimensional rod, says, $\sigma^T = -E\alpha T$, the thermal stress in the thin
plate is $1/(1 - v)$ times the thermal stress in the rod. Therefore, in the
two-dimensional thermoelastic problem, if the rise of temperature and the constraint
in two directions are the same as that in one-dimensional problem, the thermal
stress in the two-dimensional problem is $1/(1 - v)$ times the thermal stress in the
one-dimensional problem. Take $v = 0.3$ as the value of Poisson's ratio, the multiple
factor is 1.43.

8.4   Thermal Stress in Grain-Reinforced Composites

8.4.1   Research Status of Thermal Stress in Composites

Composite is a potential structural and functional material in the future due to their particular advantages. However, there exist essential or even fatal defects in composite material, one of which is the thermal stress due to the misfits of elastic modulus and thermal expansion coefficient of each component. Therefore, a proper analysis and evaluation of the thermal stress in composite material is very important for fabrication and application of composite material [1–3].

Thermal stress in multiphase materials can be measured by neutron diffraction, and X-ray, etc., in experiment generally [4, 5]. Theoretical calculations of thermal stress in composite are mainly based on the single embedding model, i.e., a single grain embedding in a hole of infinite matrix [1–5]. However, it is impossible to take into account the effect of interaction among grains on thermal stress due to the reinforcement grain appeared in the model material only once in the single embedding model. This results in a special problem of thermal stress evaluation for composites containing larger volume fraction of reinforcement phase. However, the effect of the volume fraction of reinforcement grain on thermal stress is inevitable in many cases. The effect approaches to zero only if the distance between grains is three times of the grain radius, which corresponds to the volume fraction of reinforcement phase $(1/3)^3 \approx 3.7\%$. In nowadays, it is very common for composite containing over 30% volume fraction reinforcement phase. Therefore, it is actually significant to assess the effect of volume fraction of reinforcement phase on thermal stress.

In this section, a double embedding model that assesses thermal stress in composite is presented.

8.4.2   Double Embedding Model

Zheng et al. developed the double embedding model in 1995 [6].

In general, the reinforced grain is distributed into the matrix of the composite randomly, which is surrounded by other grains inevitably. The scene can be shown in Fig. 8.4. The interaction between grains is transmitted by the nearby matrix. Figure 8.5 is schematic picture of the double embedding model to show the composite material as a first-order approximation, in which the spherical reinforced grain is embedded into hollow matrix ball and the latter is inlet into the composite material. In Fig. 8.5, a, R and b represent the radius of a reinforced grain, the outer radius of the hollow matrix in stress-free state, and the interface radius of elastic and elastoplastic zones, respectively.

Fig. 8.4 Grain-reinforced composite, ⊘ grains

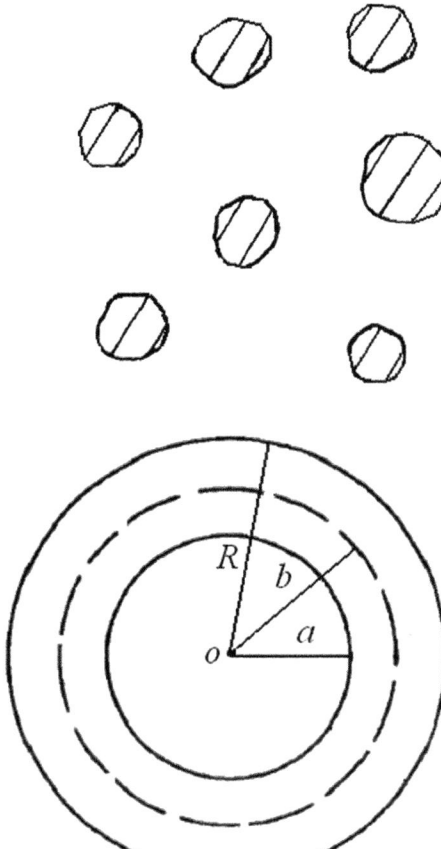

Fig. 8.5 Double embedding model

(1) *Thermal Elasticity*

Let E_0, E_1, E_2, v_0, v_1, v_2, α_0, α_1, and α_2 be the elasticity modulus, Poisson's ratio, thermal expansion coefficient of composite, matrix and reinforcement, respectively; T_0 be the temperature in stress-free state; $t = T - T_0$ be the temperature change; r be the radius of spherical grain from the center; P_0 and P_1 be the pressure at $r = R$ and a, respectively; U_{0R} and U_{1R} be the radial displacements of composite and matrix at $r = R$; U_{1a} and U_{2a} be the radial displacements of matrix and spherical grain at $r = a$ due to temperature change. According to thermal elasticity theory [7, 8], there exist

$$U_{0R} = \alpha_0 tR - \frac{P_0(1 + v_0)R}{2E_0}, \qquad (8.29)$$

$$U_{1R} = \alpha_1 tR + \frac{(1+v_1)R}{E_1} \cdot \left[\frac{\frac{a^3}{2R^3} + \frac{1-2v_1}{1+v_1}}{1 - \frac{a^3}{R^3}} \cdot P_0 - \frac{\frac{1}{2} + \frac{1-2v_1}{1+v_1}}{\frac{R^3}{a^3} - 1} \cdot P_1 \right], \qquad (8.30)$$

$$U_{1a} = \alpha_1 ta + \frac{(1+v_1)a}{E_1} \cdot \left[\frac{\frac{1}{2} + \frac{1-2v_1}{1+v_1}}{1 - \frac{a^3}{R^3}} \cdot P_0 - \frac{\frac{R^3}{2a^3} + \frac{1-2v_1}{1+v_1}}{\frac{R^3}{a^3} - 1} \cdot P_1 \right], \qquad (8.31)$$

$$U_{2a} = \alpha_2 ta + \frac{P_1(1 - 2v_1)a}{E_2}, \qquad (8.32)$$

Let $\eta = a^3/R^3$ be equal to the volume fraction of reinforcement phase in the composite, which ensures self-consistent of the solution for this problem. Furthermore, it assumed $\xi = (1 - 2v_1)/(1 + v_1)$, $U_{0R} = U_{1R}$, $U_{1a} = U_{2a}$, then it derives

$$P_0 = \left\{ (\alpha_1 - \alpha_0)t + \frac{1}{4G_1} \cdot \frac{1+2\xi}{\eta^{-1} - 1} \cdot \frac{(\alpha_2 - \alpha_0)t}{\frac{1}{4G_1} + \frac{1}{3K_2}} \right\}$$
$$+ \left\{ \frac{1+2\xi}{4G_1} \cdot \frac{\frac{1}{4G_1} - \frac{1}{4G_0}}{\left(\frac{1}{4G_1} + \frac{1}{3K_2}\right) \cdot (\eta^{-1} - 1)} - \frac{\eta + 2\xi}{4G_1(1 - \eta)} - \frac{1}{4G_0} \right\}, \qquad (8.33)$$

$$P_1 = \left\{ (\alpha_1 - \alpha_2)t + \frac{(1+2\xi)(\alpha_2 - \alpha_0)t}{4G_1(1 - \eta)\left(\frac{1}{4G_1} - \frac{1}{4G_0}\right)} \right\}$$
$$+ \left\{ \frac{1}{3K_2} + \frac{1+2\xi\eta}{4G_1(1+v_1)} - \frac{1 - 2\xi}{4G_1(1 - \eta)} \cdot \frac{\frac{1}{4G_1} + \frac{1}{3K_2}}{\frac{1}{4G_1} - \frac{1}{4G_0}} \right\}, \qquad (8.34)$$

in which $G_0 = \frac{E_0}{2(1+v_0)}$, $G_1 = \frac{E_1}{2(1+v_1)}$, $K_2 = \frac{E_2}{3(1-2v_2)}$ represent the relevant shear modulus and volume modulus, respectively.

Because E_2, E_1, E_0, α_2, α_1, and α_0 represent the elastic modulus and thermal expansion coefficient of harder phase, softer phase and composite materials, respectively, it holds $E_1 < E_0 < E_2$, $\alpha_1 < \alpha_0 < \alpha_2$, $G_1 < G_0 < G_2$. From Eq. (8.34) and the discussion below, $P_1 < 0$ and $0 < P_0$ for $t < 0$ in the range of $0 < \eta < 0.9$.

From elasticity mechanics theory, the distribution of thermal stress in different regions of the composite model is [7, 8]

$$\sigma_r = \sigma_\theta = \sigma_\varphi = P_1 \quad (r < a), \qquad (8.35)$$

$$\sigma_r = \frac{\frac{R^3}{r^3} - 1}{\eta^{-1} - 1} \cdot P_1 + \frac{1 - \frac{a^3}{r^3}}{1 - \eta} \cdot P_0 \quad (a < r < R), \qquad (8.36)$$

$$\sigma_\theta = \sigma_\varphi = -\frac{\frac{R^3}{2r^3} + 1}{\eta^{-1} - 1} \cdot P_1 + \frac{1 + \frac{a^3}{2r^3}}{1 - \eta} \cdot P_0 \quad (a < r < R), \tag{8.37}$$

$$\sigma_r = \frac{R^3}{r^3} \cdot P_0 \quad (R < r), \tag{8.38}$$

$$\sigma_\theta = \sigma_\varphi = -\frac{R^3}{2r^3} \cdot P_0 \quad (R < r), \tag{8.39}$$

The above formulae indicate that the reinforcement ($r < a$) bears compressive stress mainly under condition of $t < 0$; there exists tensile hydrostatic stress in the hollow matrix sphere zone ($a < r < R$), $\frac{1}{3}(\sigma_r + \sigma_\theta + \sigma_\varphi) = -\frac{\eta P_1}{1 - \eta} + \frac{P_0}{1 - \eta}$; the hydrostatics stress in the composite material zone ($R < r$) is zero in spite of complex deviatoric stresses.

The equivalent stress can be obtained with Eqs. (8.35) and (8.36) [8, 9]

$$\sigma_e = 0 \quad (r < a), \tag{8.40}$$

$$\sigma_e = \frac{3}{2} \cdot \frac{\frac{a^3}{r^3}}{1 - \eta} \cdot |P_1 - P_0| \quad (a < r < R), \tag{8.41}$$

$$\sigma_e = \frac{3}{2} \cdot \frac{R^3}{r^3} \cdot |P_0| \quad (R < r), \tag{8.42}$$

The misfits of the elasticity modulus and the thermal expansion coefficient at the interface of the matrix and reinforcement are the most significant. Therefore, the equivalent stress at $r = a$ is the largest. Moreover, according to Eq. (8.41), there is

$$\sigma_e|_{r=a} = \frac{3}{2} \cdot \frac{1}{1 - \eta} \cdot |P_1 - P_0| \tag{8.43}$$

As the equivalent stress described by Eq. (8.43) reaches to the yielding strength of the matrix, σ_{ys}, the matrix yields at $r = a$ first; and therefore,

$$\sigma_{ys} = \sigma_e|_{r=a} = \frac{3}{2} \cdot \frac{1}{1 - \eta} \cdot |P_1 - P_0| \tag{8.44}$$

The initial temperature difference t_{p0} for the matrix plastic yielding at $r = a$ can be obtained by means of Eq. (8.44)

(2) *Thermal Elastoplasticity*

As the temperature difference t reaches its initial critical value t_{p0}, the matrix at $r = a$ yields first. The yielding region propagates into the matrix further as the temperature difference increases. The problem of thermal stress under this condition now becomes an elastoplastic one.

As shown in Fig. 8.5, under condition of temperature difference $t > t_{p0}$, there will be a plastic yielding region around the grain, one can assume that the interface of elastic and plastic regions is at $r = b$, the radial stresses at $r = b$, $r = a$, and R are P_b, P_1, and P_0, respectively, from elastic mechanics [7, 8], it yields

$$\sigma_r = \frac{\frac{R^3}{r^3} - 1}{\frac{R^3}{b^3} - 1} \cdot P_b + \frac{1 - \frac{b^3}{r^3}}{1 - \frac{b^3}{R^3}} \cdot P_0 \quad (b < r < R), \tag{8.45}$$

$$\sigma_\theta = \sigma_\varphi = -\frac{\frac{R^3}{2r^3} + 1}{\frac{R^3}{b^3} - 1} \cdot P_b + \frac{1 + \frac{b^3}{r^3}}{1 - \frac{b^3}{r^3}} \cdot P_0 \quad (b < r < R), \tag{8.46}$$

$$u_r = \frac{-(1 + v_1)r}{E_1} \cdot \left[\frac{\frac{R^3}{2r^3} + \frac{1 - 2v_1}{1 + v_1}}{\frac{R^3}{b^3} - 1} \cdot P_b - \frac{\frac{b^3}{2r^3} + \frac{1 - 2v_1}{1 + v_1}}{1 - \frac{b^3}{R^3}} \cdot P_0 \right] + \alpha_1 t \cdot r \quad (b < r < R), \tag{8.47}$$

In the case of the matrix being an elastic–perfect plastic one, the solutions from elastoplastic mechanics are [7–9]

$$\sigma_r = 2\sigma_{ys} \ln \frac{r}{a} + P_1 \quad (a < r < b), \tag{8.48}$$

$$\sigma_\theta = \sigma_\varphi = 2\sigma_{ys} \ln \frac{r}{a} + P_1 + \sigma_{ys} \quad (a < r < b), \tag{8.49}$$

$$u_r = \frac{c_2 \sigma_{ys}}{3r^2} + 2 \cdot \frac{1 - 2v_1}{E_1} \cdot \sigma_s \cdot r \cdot \ln \frac{r}{a} + \frac{1 - 2v_1}{E_1} \cdot P_1 r + \alpha_1 t \cdot r \quad (a < r < b), \tag{8.50}$$

By using continuous conditions for displacement and stress at $r = b$, $\sigma_r|_{b-0} = \sigma_r|_{b+0}$, $U_r|_{b-0} = U_r|_{b+0}$, $\sigma_r|_b - \sigma_\theta|_b = \sigma_{ys}$, it derives

$$-\frac{(1 + v_1)b}{E_1} \cdot \left[\frac{\frac{R^3}{2b^3} + \frac{1 - 2v_1}{1 + v_1}}{\frac{R^3}{b^3} - 1} \cdot P_b - \frac{\frac{1}{2} + \frac{1 - 2v_1}{1 + v_1}}{1 - \frac{b^3}{R^3}} \cdot P_0 \right]$$
$$= \frac{C_2 \sigma_{ys}}{3b^2} + \frac{1 - 2v_1}{E_1} \cdot \sigma_{ys} \cdot b \cdot \ln \frac{b}{a} + \frac{1 - 2v_1}{E_1} \cdot P_1 \cdot b \tag{8.51}$$

$$2\sigma_{ys} \cdot \ln \frac{b}{a} + P_1 = P_b \tag{8.52}$$

$$\frac{3R^3}{2(R^3 - b^3)} \cdot |P_b - P_0| = \sigma_{ys} \tag{8.53}$$

The radial continuous conditions for displacement at $r = R$ and $r = a$ are

$$
-\frac{(1+v_0)R}{E_0} \cdot \frac{P_0}{2} + \alpha_0 tR
$$
$$
= -\frac{(1+v_2)R}{E_1} \cdot \left(\frac{\frac{1}{2} + \frac{1-2v_1}{1+v_1}}{\frac{R^3}{b^3} - 1} \cdot P_b - \frac{\frac{b^3}{R^3} + \frac{1-2v_1}{1+v_1}}{1 - \frac{b^3}{R^3}} \cdot P_0 \right) + \alpha_1 tR,
\tag{8.54}
$$

$$
\frac{(1-2v_2)ap_1}{E_2} + \alpha_2 ta = \frac{c_2 \sigma_{ys}}{3a^2} + \frac{1-2v_1}{E_1} \cdot P_1 a + \alpha_1 ta,
\tag{8.55}
$$

The above formulae show,

$$
P_1 = \frac{(\alpha_2 - \alpha_1)t - \frac{(1-v_1)}{3E_1} \cdot \frac{b^3}{a^3} \cdot \sigma_{ys}}{\frac{1-2v_1}{E_1} - \frac{1-2v_2}{E_2}},
\tag{8.56}
$$

$$
P_b = P_1 + 2\sigma_{ys} \cdot \ln\frac{b}{a},
\tag{8.57}
$$

$$
P_0 = P_b + \frac{2\sigma_{ys}}{3} \cdot \left(1 - \frac{b^3}{R^3} \right),
\tag{8.58}
$$

$$
C_2 = \frac{(1-v_1)b^3}{E_1},
\tag{8.59}
$$

Here, b can be obtained by solving the following equation,

$$
\frac{\sigma_{ys}}{3E_1} \cdot \left[2(1-2v_1) + (1+v_1) \cdot \frac{b^3}{R^3} \right] + (\alpha_1 - \alpha_0)t
$$
$$
= -\frac{2\sigma_{ys}}{E_1} \cdot \left[(1-2v_1)\ln\frac{b}{a} + \frac{E_1(1+v_0)}{2E_0} \cdot \ln\frac{b}{a} + \frac{(1+v_0)}{6E_0} \cdot \left(1 - \frac{b^3}{R^3} \right) \right]
$$
$$
+ \frac{(\alpha_1 - \alpha_2)t + \frac{(1-v_1)}{3E_1} \cdot \frac{b^3}{a^3} \cdot \sigma_{ys}}{\frac{1-2v_1}{E_1} - \frac{1-2v_2}{E_2}} \cdot \left[\frac{(1-2v_1)}{E_1} + \frac{(1+v_0)}{2E_0} \right],
$$
$$
\tag{8.60}
$$

As $b = R$, the matrix yields plastically overall. Under such condition, the temperature difference t_p is obtained,

$$t_p = \frac{-\sigma_{ys}}{E_1}$$

$$\cdot \left\{ (1 - v_1) + 2(1 - 2v_1) \ln \frac{R}{a} + \frac{E_1(1 + v_0)}{E_0} \cdot \ln \frac{R}{a} + \frac{\frac{(1-v_1)}{3} \left[\frac{(1-2v_2)}{E_1} + \frac{(1+v_0)}{2E_0} \right]}{\left(\frac{1-2v_1}{E_1} - \frac{1-2v_2}{E_2} \right) \cdot \frac{a^3}{R^3}} \right\}$$

$$+ \left[(\alpha_1 - \alpha_0) - \frac{\left(\frac{1-2v_1}{E_1} + \frac{1+v_0}{2E_0} \right)(\alpha_1 - \alpha_2)}{\frac{1-2v_1}{E_1} - \frac{1-2v_2}{E_2}} \right]$$

$$(8.61)$$

(3) The Calculation of E_0 and α_0

Zheng et al. proposed an expression for assessing the effective elasticity modulus of multiphase composite [10],

$$\sum_{i=1}^{n} \frac{\eta_i}{\frac{1-2v_i}{E_i} + \frac{1+v_0}{2E_0}} = \frac{2E_0}{3(1 - v_0)}, \tag{8.62}$$

where n is the number of components in the composite; E_i, v_i, and η_i represent the elastic modulus, the Poisson's ratio, and the volume fraction of i-th component, respectively; E_0 and v_0 are the effective elastic modulus and Poisson's ratio of the composite. For two-component system, η can be the volume fraction of the reinforced phase, then Eq. (8.62) becomes

$$\frac{\eta}{\frac{1-2v_2}{E_2} + \frac{1+v_0}{2E_0}} + \frac{1 - \eta}{\frac{1-2v_1}{E_1} + \frac{1+v_0}{2E_0}} = \frac{2E_0}{3(1 - v_0)} \tag{8.63}$$

Proper results were obtained by assuming $v_1 \approx v_2 \approx v_0 \approx 0.3$ for ceramics reinforced metal-matrix composite, and $v_1 \approx v_2 \approx v_0 \approx 0.2$ for multiphases ceramics [10].

Turner's formula can be used to evaluate α_0 [11],

$$\alpha_0 = \frac{\sum_{i=0}^{n} \alpha_i E_i \eta_i}{\sum_{i=1}^{n} E_i \eta_i} \tag{8.64}$$

For two-component system, Eq. (8.64) becomes

$$\alpha_0 = \frac{\alpha_1 E_1(1 - \eta) + \alpha_2 E_2 \eta}{E_1(1 - \eta) + E_2 \eta} \tag{8.65}$$

8.4.3 Discussion

According to the single embedding model [4], the thermal stress at interface of grain and matrix within elastic range can be written as

$$P_1 = \frac{(\alpha_1 - \alpha_2)t}{\left(\frac{1+v_1}{2E_1} + \frac{1-2v_2}{E_2}\right)} \tag{8.66}$$

Equation (8.66) indicates that the thermal stress due to temperature change is independent of volume fraction η of reinforcement phase, which is inconsistent with practical condition and the predictions of double embedding model (see Eq. (8.34)). In fact, Eq. (8.34) could reduce to Eq. (8.66) if η approaches zero, i.e., the single embedding model is a special case of double embedding model for η approaching zero in practice, and this indicates that single embedding model is only suitable for analyzing thermal stress of composite containing smaller volume fraction of reinforcement phase.

Take SiC grain-reinforced aluminum composite as an example, the physical coefficients are: $\alpha_1 = 22.5 \times 10^{-6}$ K^{-1}, $E_1 = 70$ GPa, $\alpha_2 = 3.3 \times 10^{-6}$ K^{-1}, $E_2 = 200$ GPa. Assuming $v_1 \approx v_2 \approx v_0 \approx 0.3$, the variation of P_1/t with η can be obtained from Eqs. (8.34), (8.63), (8.65), and (8.66), as is shown in Fig. 8.6. In Fig. 8.6, the dotted line is the predicted result by using Eqs. (8.34), (8.63), and (8.65), and the solid line is that by using Eq. (8.66). P_1/t predicted by Eq. (8.34) decreases with increasing η, especially in the range of $0.9 < \eta < 1$, P_1/t becomes negative with very small values, as a result of the interaction of grains on the thermal stress at the interface of grain and matrix. Practically, composite materials as a whole become firmer and hardly deform with the increasing volume fraction η of reinforced phase. Therefore, the matrix in zone $a < r < R$ in Fig. 8.6 bears stronger constraint from the composite in zone of $r > R$ due to temperature change with greater η. This

Fig. 8.6 P_1/t versus η for SiC–Al composite

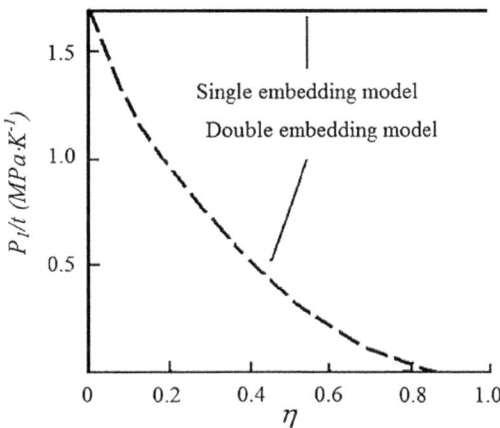

Single embedding model

Double embedding model

P_1/t (MPa·K^{-1})

1.5

1.0

0.5

0   0.2   0.4   0.6   0.8   1.0

η

Fig. 8.7 P_0/t versus η for
SiC–Al composite

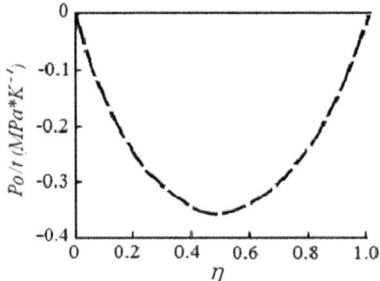

Fig. 8.8 t_{p0} versus η for
SiC–Al composite

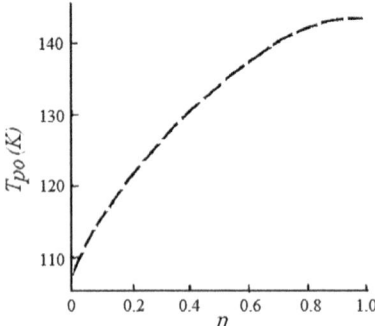

will change the thermal stress at the interface of grain and matrix. However, as to the single embedding model, Eq. (8.66) does not derive this result. Figure 8.7 shows the relationship between P_0/t and η in the range of $0 < \eta < 1$. P_0/t is always negative; therefore for $t < 0$, P_0 is a tensile stress. Substituting all these results into Eqs. (8.35) through (8.39), it obtains that the reinforced grain ($r < a$) bears compressive hydrostatic stress in $0 < \eta < 0.9$ during cooling; matrix ($a < r < R$) stands for tensile hydrostatic stress $\frac{\sigma_r + \sigma_\theta + \sigma_\varphi}{3} = -\frac{\eta P_1}{1-\eta} + \frac{P_0}{1-\eta}$; composite materials zone ($R < r$) is hydrostatic stress-free.

By substituting the plastic yielding strength of aluminum $\sigma_{ys} = 274$ MPa into Eqs. (8.33), (8.34), (8.63) and (8.65), and Eqs. (8.61), (8.62) and (8.65) [4], it obtains the temperature difference of initial yielding at the interface of grain and matrix and the overall yielding of the matrix, respectively. Figure 8.8 shows the variations in initial temperature difference t_{p0} at the interface of SiC grain and Al matrix with respect to η. It shows that t_{p0} increases with η rising, which is the inevitable result of the interaction of grains. Figure 8.9 represents the variations of temperature t_p of the Al matrix plastic yielding overall with respect to η. t_p decreases as η increases, because the volume fraction $(1 - \eta)$ of Al decreases as the volume fraction of SiC η increases. At the same time, the number of the core of the initial yielding increases, thus once the interface of matrix and grain yields, it propagates into the matrix rapidly; and therefore, a small change in temperature will induce the total matrix Al yield plastically.

Fig. 8.9 t_p versus η for
SiC–Al composite

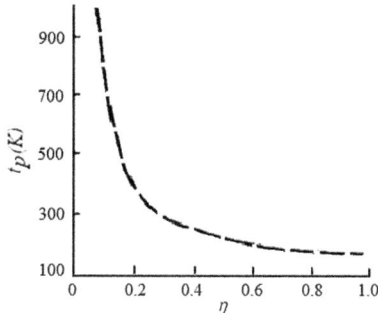

8.4.4   Summary

In nowadays, grain-reinforced metal-matrix composites and multiphase ceramics
are quite common. Appropriate assessment of thermal stress in composite is
indispensable to the research and application of composite. The double embedding
model is superior to other models in thermal stress assessment, which could consider the interaction among grains.

References

1. Lee EU (1992) Thermal stress and strain in a metal matrix composite with a spherical
   reinforcement particle. Metall Trans 23A:2205–2210
2. Takei T, Hatta H, Taya M (1991) Thermal expansion behavior of particulate-filled
   composites. Mater Sci Eng A 131:133–143
3. Povik GL, Needleman A, Nutt SR (1990) An analysis of residual stress formation in
   whisker-reinforced Al–SiC. Mater Sci Eng A 125:129–140
4. Ledbetter HM, Austin MW (1987) Internal strain(stress) in an SiC–Al particle-reinforced
   composite: an X-ray diffraction study. Mater Sci Eng 89:53–61
5. Davis LC, Allison JE (1993) Residual stresses and their effects on deformation in
   particle-reinforced metal-matrix composites. Metall Trans A 24:2487–2496
6. Zheng M, Jin Z, Hao H (1995) A new model for analyzing thermal stress in granular
   composite. Sci China (Ser A) 38(6):739–748
7. Xu Z (1988) Elastic mechanics, 2nd edn, vol 1. Advanced Educational Press, Beijing, China,
   pp 277–279
8. Xu B, Huang Y, Liu X et al (1984) Elastoplastic mechanics and its application (in Chinese).
   Mechanical Industrial Press, Beijing, China, pp 131–152
9. Kachanov LM (1983) Introduction to plasticity (in Chinese), 2nd edn. People's Educational
   Press, Beijing, China, pp 93–97, 101–115
10. Zheng M, Zheng X (1991) Expressions for predicting the effective elasticity modulus of
    materials reinforced by 2nd-phase grains. Metall Trans A 22:507–511
11. Guan Z, Zhang Z, Jiao J (1992) Physical properties of inorganic materials (in Chinese).
    Tsinghua University Press, Beijing, China, pp 119–130

Chapter 9
Fatigue Behavior of Highly Ductile Materials

Abstract The general description of fatigue problems is given first, and then the uniform fatigue life equation for both low-cycle fatigue and high-cycle fatigue conditions and its improvement are presented, which include the mean stress effect.

9.1 General Description of Fatigue Problems

In general, every component of machines, vehicles, and structures is subjected to repeating loads inevitably, which results in a cyclic stress or load acting to the components, and microscopic physical damage forms within the materials involved gradually.

Even the stress is well below the yielding strength of a given material, and the microscopic scaled damage could accumulate with continued cycling until it develops into a small crack, or macroscopic failure of the component. This process of gradual damage and failure corresponding to the cyclic loading is called *fatigue*.

Mechanical failure due to fatigue has been the subject in engineering for nearly 200 years. Albert W. A. J. was one of the early researchers, who tested mine hoist chains under cyclic loading in Germany around 1828. Early in 1839, the term *fatigue* was used by Poncelet J. V. in a book on mechanics. Fatigue was further discussed and studied in the mid-1800s by a number of peoples in many countries to dealing with the problem of failures of components such as stagecoach and railway axles, shafts, gears, beams, and bridge girders.

In 1850s, the failures of railway axle motivated Wŏhler August to work on this topic in Germany specifically. He attempted to design a strategy to avoid fatigue failure, thus various loads, such as bending, torsion, and axial loadings, were applied to irons, steels, and other metals to conduct experiment. His results indicated that fatigue life was influenced by both cyclic stresses and the accompanying steady (mean) stresses. Furthermore, Gerber and Goodman's studies attempted to predict the effects of mean stress.

Fatigue failure has attracted major concern in engineering design continuously. It was estimated that 80% of these costs involve cyclic loading and fatigue in the

© Springer Nature Singapore Pte Ltd. 2019

M. Zheng et al., *Elastoplastic Behavior of Highly Ductile Materials*,

https://doi.org/10.1007/978-981-15-0906-3_9

economic costs of fracture. Consequently, the account of about 3% in the gross national product (GNP) of US economy is lost due to the annual cost of fatigue of materials, and a similar percentage is estimated for other industrial nations. Fatigue failure is involved in many fields ranging from ground vehicles, rail vehicles, bridges, cranes, power plant equipment, offshore oil well structures, to aircraft of all types, almost everywhere, including everyday household items, toys, and sports equipment.

At present, there are three major approaches to the analysis and design against fatigue failures. The traditional *stress-based approach* or *high-cycle fatigue* was developed in 1955. Here, the analysis is based on the nominal (average) stresses in the relevant region of the engineering component. Another approach is the *strain-based approach*, which involves higher cyclic strain range and therefore less fatigue life ($<10^5$), so it is also called low-cycle fatigue. More detailed analysis of the localized yielding that may form at stress raisers during cyclic loading is employed in general. Finally, it is the *fracture mechanics approach*, which is specifically to treat growing cracks with fracture mechanics methods.

Fatigue is a complex physical process due to the lacking of the interactions of various geometric position and material parameters particularly. The difficulties and complexities in modeling the fatigue crack growth process have been described in [1, 2]. In general, many fatigue tests on materials involve cycling between maximum and minimum stress levels that are constant, which is called *constant amplitude stressing* and is illustrated in Fig. 9.1.

The *stress range* is defined by the difference of the maximum and the minimum values of the applied stress, i.e., $\Delta\sigma = \sigma_{max} - \sigma_{min}$. The average value of the maximum and minimum stresses is named as the *mean stress*, $\sigma_m = 0.5 \cdot (\sigma_{max} + \sigma_{min})$. The mean stress may be zero, as in Fig. 9.1a, but it is not zero in general, as in (b). *The stress amplitude* (someone calls it as *alternating stress*) is defined as the half range of the stress range, i.e., $\sigma_a = 0.5 \cdot \Delta\sigma = 0.5 \cdot (\sigma_{max} - \sigma_{min})$.

Furthermore, the following two ratios are used sometimes: $R = \sigma_{min}/\sigma_{max}$, and $A = \sigma_a/\sigma_m$; R is called the *stress ratio* and A the *amplitude ratio*, respectively. Then, some simple relationships can be derived from the preceding equations: $\sigma_a = \sigma_{max}(1 - R)/2$, $\sigma_m = \sigma_{max}(1 + R)/2$, $R = (1 - A)/(1 + A)$, $A = (1 - R)/(1 + R)$.

If an engineering component or a test specimen is subjected to an adequately severe cyclic stress, there will be an initiation of fatigue crack or other damage gradually, which results in a complete failure of the member finally. The cyclic number of failure depends upon the stress level.

A *stress–life curve* could be plotted with the results of the tests from a number of different stress levels, which is called an *S-N curve*. The amplitude of stress or nominal stress, σ_a or S_a, and the cyclic number of failure N_f are usually taken as the y-axis and x-axis, respectively, as shown in Fig. 9.2.

A group of such fatigue tests gives an *S-N curve*, which may be run all at zero mean stress, or all at some specific value of nonzero mean stress, σ_m. The common *S-N curve* is for a constant value of the stress ratio, R. In general, stress is usually plotted as amplitude, but $\Delta\sigma$ or σ_{max} is sometimes plotted instead.

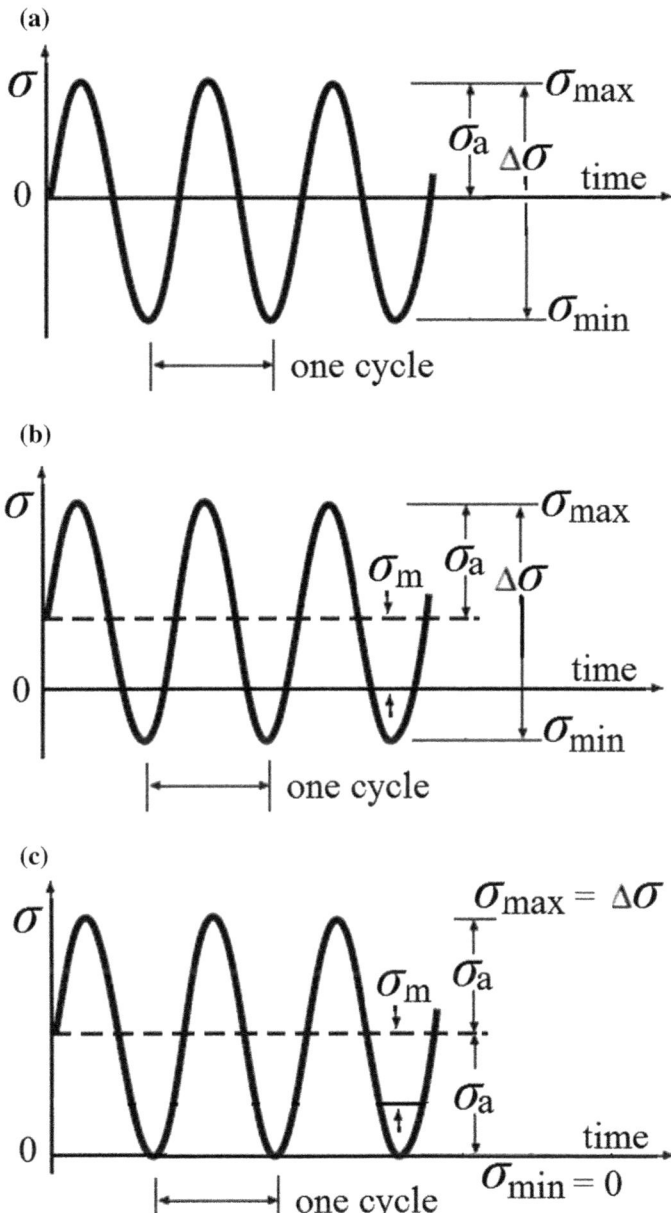

Fig. 9.1 Typically cyclic load and its nomenclature. **a** Complete reversal cyclic stress, $\sigma_m = 0$; **b** nonzero mean stress σ_m; and **c** zero-to-tensile stress, $\sigma_{min} = 0$

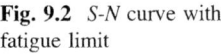

Fig. 9.2 *S-N* curve with
fatigue limit

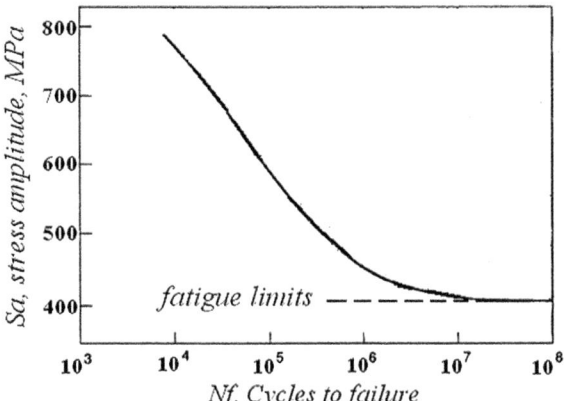

The cyclic number of failure changes rapidly with stress level and may even cover several orders of magnitude. So, the cycle numbers are usually plotted on a logarithmic scale as well, which is illustrated in Fig. 9.2 with logarithmic scale of N_f.

Usually, there appears a distinct stress level, below which fatigue failure does not occur under ordinary conditions in some metals, such as plain-carbon and low-alloy steels for example. This is illustrated in Fig. 9.2, in which the *S-N* curve tends to become flat and to approach asymptotically the stress amplitude labeled by S_e. Such lower limiting stress amplitude is called *fatigue limits* or *endurance limit*. For a smooth surface finished specimen without notch testing, it is denoted by σ_e, and it is considered as a material property generally.

The term *fatigue strength* is used to specify a stress amplitude value from an *S-N* curve at a particular life as defined. Hence, the *fatigue strength* at 10^5 cyclic number is simply the stress amplitude which corresponds to $N_f = 10^5$. The terms *high-cycle fatigue* and *low-cycle fatigue* with *S-N* curves indicate the situations of long fatigue life (sufficiently low stress) and short fatigue life (sufficiently high stress), respectively.

In the viewpoint of practical application, the interest lies to relate the fatigue life of structural component to the allowable range of cyclic loading [1, 2]. Semi-empirical methods are employed frequently to get useful results. The difference of metal fail under monotonically rising load and cyclic action is quite significance [3–10]. The former involves obviously plastic deformation which dissipates the ductility of metals due to the accumulation of micro damage gradually [3]. The latter is more complex [4, 5], the fatigue life of metal is influenced by the temperature in special manner, and it cannot always be extended by the increase in strength, or ductility solely. That is, the effect of strength and ductility on fatigue is more complicated.

9.2 Fatigue Crack Initiation

Since the early work of Coffin and Manson in the 1950s and the 1960s, the phenomenon related to low-cycle fatigue has received much attention, and the feature of low-cycle fatigue is different from that of high-cycle fatigue. The high-cycle fatigue was considered as a result of an elastic phenomenon in macroscale. While, in the process of a low-cycle fatigue, a macroscopic plastic deformation is involved in every cycle.

Under low-cycle fatigue, it fails in a small number of cycles, even 10^3 cycles or less. Small crack nucleates immediately generally. Low-cycle fatigue is relevant to the component that underwent a small numbers of cycles in the loading service. In view of the high stress level, final failure will occur when the crack is still small in size. If stress level is less than the fatigue limit, the structure could be stable.

Coffin and Manson independently found that the fatigue life can be represented by a function of the strain amplitude in the case of low-cycle fatigue, ε_a, and it shows a linear relation in a double logarithmic scale. The correlation could be expressed as follows:

$$\varepsilon_a \cdot N^\beta = \text{constant } C, \quad \text{or } \varepsilon_a = C \cdot N^{-\beta} \tag{9.1}$$

Equation (9.1) is known as the Coffin–Manson relation, and the exponent β is in the order of -0.5 usually.

However, in case of high-cycle fatigue, the Coffin–Manson relation is not valid remarkably. The lower horizontal asymptote, i.e., the fatigue limit, is not included in Eq. (9.1), see Fig. 9.2. However, by using a similar exponential relation, Manson and Hirschberg also included the elastic strain amplitude into the fatigue life, which results in

$$\varepsilon_{a,\text{total}} = \varepsilon_{a,\text{pl}} + \varepsilon_{a,\text{el}} = C_1 N^{-\beta 1} + C_2 N^{-\beta 2}, \tag{9.2}$$

Figure 9.3 shows a general fatigue result for stainless steel. The curve for $\varepsilon_{a,\text{ total}}$ becomes a more horizontal direction at high endurances where the elastic strain amplitude predominates. There are four material constants in Eq. (9.2). However, the question remains how useful it can be for the fatigue life prediction problem.

9.2.1 Mean Stress Effect

The mean stress effect is an important factor [11–14], and it needs to be applied in fatigue life assessment. It is useful to see the curves of strain–life, in which the actual value of mean stress is indicated, as shown in Fig. 9.4 for an alloy steel.

In order to reveal mean stress effect, Goodman, Gerber, Morrow, SWT, and Walker proposed different methods, and their approaches are with different

Fig. 9.3 Sketch of
$\Delta\varepsilon_{\text{total}} = \Delta\varepsilon_{\text{el}} + \Delta\varepsilon_{\text{pl}}$

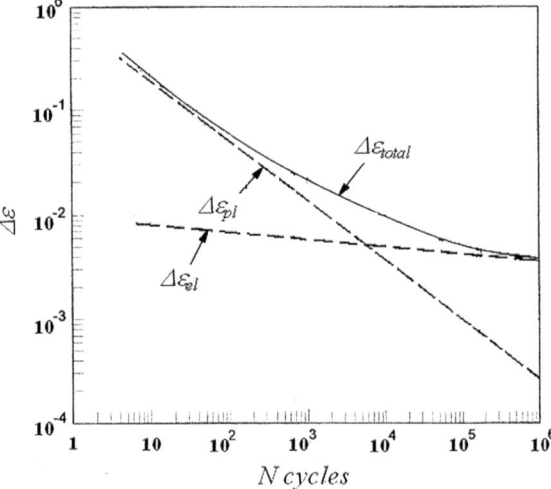

Fig. 9.4 Mean stress effect
on the strain–life of an alloy
steel

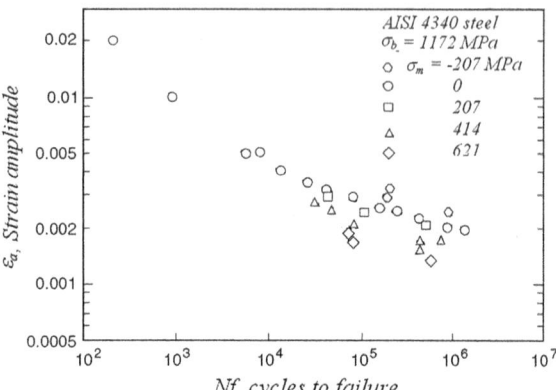

accuracies. As a result, the general recognized and commonly applied methods to strain–life assessment are those proposed by Morrow, SWT, and Walker [12–14].

Morrow J. proposed an approach to reflect the mean stress effect in 1968 [11]. If the stress amplitude for the zero mean stress ($\sigma_m = 0$) is designated by σ_{ar}, while stress amplitude for the nonzero mean stress ($\sigma_m \neq 0$) is σ_a, then there is an appropriate relationship on a constant life diagram,

$$\frac{\sigma_a}{\sigma_{ar}} + \frac{\sigma_m}{\sigma'_f} = 1, \tag{9.3}$$

In Eq. 9.3, σ'_f is the fracture strength of metal in tensile test, which is often higher than σ_b for ductile metals.

An alternate form of Eq. (9.3) is

$$\frac{\sigma_a}{\sigma_{ar}} = 1 - \frac{\sigma_m}{\sigma'_f}. \tag{9.4}$$

and

$$\sigma_{ar} = \sigma_a \Big/ \left(1 - \frac{\sigma_m}{\sigma'_f}\right). \tag{9.5}$$

In Eq. (9.5), σ_{ar} and σ_a express the stress range for the zero mean stress ($\sigma_m = 0$) and the stress range for the nonzero mean stress ($\sigma_m \neq 0$), respectively.

While Smith, Watson, and Topper (SWT) proposed that the life is determined by the geometric mean of the maximum stress and the stress amplitude [12–14],

$$\sigma_{ar} = (\sigma_{max}\sigma_a)^{1/2}, \tag{9.6}$$

An alternate form of Eq. (9.6) is

$$\sigma_{ar} = \sigma_a[2/(1-R)]^{1/2}, \sigma_{ar} = \sigma_{max}[(1-R)/2]^{1/2} \tag{9.7}$$

However, Walker's equation involves an additional parameter of material property, γ, which is

$$\sigma_{ar} = \sigma_{max}^{1-\gamma}\sigma_a^{\gamma}, \sigma_{ar} = \sigma_a[2/(1-R)]^{1-\gamma}, \quad \sigma_{ar} = \sigma_{max}[(1-R)/2]^{\gamma}. \tag{9.8}$$

If $\gamma = 0.5$, Eq. (9.8) reduces to Eq. (9.7), thus Walker equation emerges as SWT's formula. The quantity $(1 - \gamma)$ can be considered as a measure of the sensitivity of material to mean stress [14].

Comparatively, SWT's method has the characteristics of simplicity.

9.2.2   An Uniform Fatigue Life Equation for Both Low-Cycle Fatigue and High-Cycle Fatigue Conditions

As to high-cycle fatigue, the Coffin–Manson relation is not valid directly. The lower horizontal asymptote, i.e., the fatigue limit, is not covered by Eq. (9.1). An alternate model of fatigue damage accumulation is proposed by Zheng et al. [4–8]; the *fatigue* has been attributed to the *local damage of plastic deformation* in metals. A hypothetical "fatigue element" was introduced to study fatigue crack initiation (FCI) at a notch [7].

The fatigue failure process of metal can be divided into three stages [7]: firstly, under cyclic stress or alternating action, fatigue cracks are generally formed on the metal surface; then, the formed crack has continued to develop; when the crack expands into the critical dimension, it leads to fracture finally. If the test piece does not form a crack under the action of a certain strain range $\Delta\varepsilon_i$, there is no fatigue failure, and the fatigue life tends to be infinite. Accordingly, it can be considered that the strain range $\Delta\varepsilon_i$ does not cause fatigue damage in the metal.

There are three ways for fatigue cracks to form: slip band opening, the interface cracking between the inclusions and the substrate or the fracture of the inclusion itself, and twin grain boundary or crystalline boundary cracking. The fatigue crack is formed by the cyclic plastic strain of the metal, regardless of the way it is formed. Therefore, the local cyclic plastic strain is the resource of fatigue damage.

In the polycrystalline metal, due to the difference in grain orientation and microstructure, plastic deformation occurs in a few grains even if the stress or strain is lower than the macroscopic elastic limit σ_e.

According to the above analysis and experimental results, the strain range $\Delta\varepsilon$ applied to the test specimen can be divided into two parts, that is, a part does not induce fatigue damage of the metal, which is the critical strain range $\Delta\varepsilon_c$, and it should be the theoretical strain fatigue limit of the metal; while the other part will induce fatigue damage of the metal, which can be called the *damage strain range* $\Delta\varepsilon_D$. It should be equal to the difference of the whole strain range subtracting the theoretical strain fatigue limit in value, i.e., $\Delta\varepsilon_D = \Delta\varepsilon - \Delta\varepsilon_c$. Therefore, the strain fatigue life N_f of a metal should be a function of the range of damage strain $\Delta\varepsilon_D$. According to the published research results, it is considered that there should be a power function relationship between N_f and $\Delta\varepsilon_D$.

$$\Delta\varepsilon_D = \Delta\varepsilon - \Delta\varepsilon_c = \varepsilon'_f \cdot N_f^b \tag{9.9}$$

Although the form of Eq. (9.9) is similar to that of Eq. (9.1), the plastic strain range $\Delta\varepsilon_p$ in Eq. (9.1) is replaced by the range of damage strain $\Delta\varepsilon_D$. In the low-cycle range where the $\Delta\varepsilon_D$ value is high and the N_f value is small, $\Delta\varepsilon_D$ approaches to $\Delta\varepsilon_p$, and thus Eq. (9.9) is very close to the Coffin–Manson formula, i.e., Eq. (9.1).

The exponent b in Eq. (9.9) is the fatigue ductility index. Coffin gives $b = -0.5$. Manson gives an average value of -0.6 for b, and it has recently been corrected to -0.56. Martin gave $b = -0.5$ by energy analysis for strain fatigue. Therefore, $b = -0.5$ could be an appropriate choice accordingly.

Furthermore, the coefficient ε'_f is a constant related to the fracture ductility of metal ε_f. In case of uniaxial tensile loading, which is equivalent to $N_f = 1/4$ cycle loading, it results in $\Delta\varepsilon_D / 2 = \varepsilon_f$. Therefore, it obtains from Eq. (9.9)

$$\varepsilon'_f = \varepsilon_f \tag{9.10}$$

Equation (9.4) is consistent with Coffin's experimental results.

By substituting b and ε'_f into Eq. (9.10) and rearrangement, *the alternate uniform formula of fatigue life* assessment is obtained [7],

$$N_i = A \cdot (\Delta\varepsilon - \Delta\varepsilon_c)^{-2} \tag{9.11}$$

$$A = \varepsilon_f^2 \tag{9.12}$$

The coefficient A can be called the *strain fatigue resistance coefficient*, and it is the material constant. It should be noted that Manson also gives a formula similar to Eq. (9.11). However, the coefficients, indices, and definitions of $\Delta\varepsilon_c$ in Manson's formula are not well described, and the constants are still needed to fit by the experimental data. Langer once proposed a strain fatigue formula similar to Eq. (9.11) based on simple assumptions as well, but the definitions and values of his constants are different from those given here.

In addition, the fatigue limit is lower than elastic limit in general, therefore it exists

$$\Delta\sigma_c = E \cdot \Delta\varepsilon_c \tag{9.13}$$

in which E is elastic modulus of material.

It can be seen from Eq. (9.11) that if the value of the coefficient A and the theoretical strain fatigue limit $\Delta\varepsilon_c$ are well estimated, the strain fatigue life curve can be obtained. In fact, the value of the coefficient A can be obtained by substituting the value of the fracture ductility ε_f of the material into Eq. (9.12).

The value of the theoretical strain fatigue limit $\Delta\varepsilon_c$ cannot be obtained from the value of the fatigue limit σ_{-1} and Eq. (9.13) directly, the experimentally determined fatigue limit σ_{-1} is at a stress ratio R of -1 by specifying the magnitude of the stress at 10^7 cycles. Therefore, the strain range corresponding to $N_f = 10^7$ cycle can be calculated under condition of σ_{-1}, i.e., $\Delta\varepsilon_{N_f = 10^7} = 2\sigma_{-1}/E$. Then, substituting these values for A, $\Delta\varepsilon_{N_f = 10^7}$, and $N_f = 10^7$ into Eq. (9.11), it obtains the assessment of theoretical strain fatigue limit $\Delta\varepsilon_c$ [7–9],

$$\Delta\varepsilon_c = 2\sigma_{-1}/E - \varepsilon_f/10^{3.5}. \tag{9.14}$$

Now, the values of the coefficient A and the theoretical strain fatigue limit $\Delta\varepsilon_c$ are well estimated with Eqs. (9.12) and (9.14), and the strain fatigue life curve can be obtained by using Eq. (9.11).

A lot of test data has been employed to check the validity of Eq. (9.11) in light of Eqs. (9.12) and (9.14) [7–9], and good agreement was obtained.

In addition, there exists empirical but simple relationship between σ_{-1} and σ_b [10], $\sigma_{-1} = 0.50 \sigma_b$ for steel with $\sigma_b < 1800$ MPa; $\sigma_{-1} = 0.35 \sigma_b$ for magnesium alloys, copper alloys, and nickel alloys; $\sigma_{-1} = 0.40 \sigma_b$ for aluminum alloys for $\sigma_b < 325$ MPa [11].

In elastic range, $\Delta S = E \cdot \Delta\varepsilon$, together with Eq. (9.13), Eq. (9.11) becomes

$$N_i = S_f \cdot (\Delta S - \Delta S_c)^{-2} \tag{9.15}$$

$$S_f = E\varepsilon_f^2/4 \tag{9.16}$$

The coefficient S_f can be called the *stress fatigue resistance coefficient*, and it is the material constant. In this condition, ΔS and ΔS_c represent the nominal stress range and the theoretical stress fatigue limit for smooth sample, respectively [7–9].

Under plastic condition by using nonlinear stress–strain relation, Holloman formula for example,

$$\sigma = K\varepsilon^n \tag{9.17}$$

in which, K presents the strain hardening coefficient, and n indicates the exponent in the uniaxial to relate stress and strain.

For complete reversal cyclic loading $R = -1$, it yields a relation

$$\Delta\varepsilon = 2[\Delta\sigma/(2K)]^{1/n} \tag{9.18}$$

Furthermore, for notched specimen, under plastic yielding condition by using the Neuber approach and together Holloman formula, the local stress and strain can be correlated with the far field quantities [7–9]. For complete reversal cyclic loading $R = -1$, it obtains [7, 9]

$$\Delta\varepsilon = 2\left[(K_t \cdot \Delta\sigma_0/2)^2 \Big/ (EK)\right]^{1/(1+n)} \tag{9.19}$$

Thus, Eq. (9.11) can thus be developed to fatigue life of cyclic stress loading for notched specimen as follows:

$$\Delta\sigma' = K_t\Delta\sigma_0/2 \tag{9.20}$$

$$N_i = C \cdot \left[(\Delta\sigma')^{2/(1+n)} - (\Delta\sigma_c')^{2/(1+n)}\right]^{-2} \tag{9.21}$$

$$\Delta\sigma_c' = (E\sigma_f\varepsilon_f)^{0.5} \cdot (\Delta\varepsilon_c/2\varepsilon_f)^{(1+n)/2} \tag{9.22}$$

$$C = 0.25(EK\varepsilon_f)^{2/(1+n)}. \tag{9.23}$$

in which C is referred to as the *fatigue strength coefficient* [4–10].

In considering the mean stress effect, by using SWT method and Eqs. (9.7) and (9.20) can be improved as [4–10]

$$\Delta\sigma' = K_t\Delta\sigma_0/[2(1-R)]^{1/2}, \Delta\sigma_c' = [2E\sigma_f\epsilon_f/(1-R)]^{1/2} \cdot (\Delta\epsilon_c/2\epsilon_f)^{(1+n)/2} \tag{9.24}$$

Thus, Eqs. (9.21), (9.22), (9.23), and (9.24) together could assess the fatigue life with mean stress effect.

In Eqs. (9.21) through (9.24), $\Delta\sigma_0$ is the nominal applied stress range, R the stress ratio, K_t the stress concentration factor, σ_f the ultimate of the metal, and ε_f and E the fracture strain and elastic modulus of the metal.

Equations (9.21) through (9.24) have been applied to analyze the fatigue behavior of several metals where the variations of stress ratio and test temperature and overload effect were included [4–10].

In the estimations of the local inelastic stresses and strains near notch, Neuber's rule has been used, which reads

$$K_t = \frac{\sigma}{\sigma_0} = \frac{\varepsilon}{\varepsilon_0} \tag{9.25}$$

In Eq. (9.18), σ and ε are the elastic–plastic local stress and strain, respectively; while σ_0 and ε_0 are the respective remote nominal stress and strain.

The elastic local stress and strain expression for definition of the stress concentration factor has been extended to the elastic–plastic case. In plastic case, plastic stress and strain concentration factors K_σ and K_ε can be written, respectively, as

$$K_\sigma = \sigma/\sigma_0, \ K_\varepsilon = \varepsilon/\varepsilon_0, \tag{9.26}$$

Combining K_σ and K_ε in Neuber's rule, it yields

$$K_t^2 = K_\sigma \cdot K_\varepsilon \tag{9.27}$$

This leads to

$$(K_t\sigma_0)^2 = E\sigma\varepsilon, \ \text{with } \sigma_0 < \sigma_{ys} \tag{9.28}$$

where σ_{ys} is the yield strength of the material.

However, Molski K. and Glinka G. pointed out that Eq. (9.28) overestimates the local strains and leads to significant errors in fatigue life prediction [15]. It was argued that the plastic energy next to the notch root is localized and does not differ from the surrounded elastic material significantly [15]. Furthermore, Molski K. and Glinka G. assumed that a state of elastic and uniaxial strain prevails at the notch root, thus the local strain energy density can be written as

$$W_\sigma = \sigma^2/2E \tag{9.29}$$

while the nominal elastic strain energy density is given by

$$W_0 = \sigma_0^2/2E \tag{9.30}$$

The elastic stress concentration factor takes the form

$$K_t = (W_\sigma/W_0)^{0.5} \tag{9.31}$$

Moreover, Molski K. and Glinka G. assumed that W_σ in Eq. (9.24) should be extended to notch root plastic condition and can be recalculated by using the nonlinear stress–strain relation in Eq. (9.11) [15]. Near the notch root, the plastic strain energy density becomes

$$W_\sigma = \int_0^\varepsilon \sigma(\varepsilon)d\varepsilon = [K \cdot \varepsilon^{(n+1)}]/(n+1)$$
$$= [\sigma/(n+1)] \cdot (\sigma/K)^{1/n} \tag{9.32}$$

while the nominal strain energy density is still given by Eq. (9.30). Substitute Eqs. (9.30) and (9.32) into Eq. (9.31), it yields

$$K_t = (W_\sigma/W_0)^{0.5} = \{(\sigma/K)^{1/n} \cdot [\sigma/(n+1)]/(\sigma_0^2/2E)\}^{0.5} \tag{9.33}$$

Equation (9.18) can be again applied in conjunction with Eq. (9.33), and it gives

$$\varepsilon = [\sigma/(K)]^{1/n} = [(K_t\sigma_0)^2/(EK)]^{1/(1+n)} \cdot [(1+n)/2]^{1/(1+n)} \tag{9.34}$$

Sih once pointed out that fatigue load sequence effects can have an obvious influence on fatigue crack growth predictions [2]. Fatigue crack growth predictions can also be obtained from the use of sole uniaxial tensile data [1, 2, 16].

9.2.3 Improvement of the Uniform Formula of Fatigue Life Assessment

By means of Molski and Glinka's energy strain density concentration near a notch, an improvement of the uniform formula of fatigue life assessment was proposed [17]. First, Eq. (9.34) will be extended to the case of full reversal loading with $R = -1$; ε, σ, and σ_0 may be replaced by $\Delta\varepsilon/2$, $\Delta\sigma/2$, and $\Delta\sigma_0/2$ for notched specimen, respectively, i.e.,

$$\Delta\varepsilon = 2(\Delta\sigma/2K)^{1/n} = 2\left[(K_t \cdot \Delta\sigma_0/2)^2/(EK)\right]^{1/(1+n)} \cdot [(1+n)/2]^{1/(1+n)} \tag{9.35}$$

Similar to Eq. (9.14), an equivalent stress range $\Delta\sigma^*$ at the notch tip can be introduced

$$\Delta\sigma^* = K_t \cdot \Delta\sigma_0/2 \tag{9.36}$$

$$\Delta\varepsilon = 2\left[(\Delta\sigma^*)^2/(EK)\right]^{1/(1+n)} \cdot \left[(1+n)/2\right]^{1/(1+n)} \tag{9.37}$$

Substituting Eq. (9.28) into Eq. (9.5), it results in

$$N_i = C_1^* \cdot \left[(\Delta\sigma^*)^{2/(1+n)} - (\Delta\sigma_c^*)^{2/(1+n)}\right]^{-2} \tag{9.38}$$

in which

$$C_1^* = 0.25\varepsilon_f^2 \cdot (EK)^{2/(1+n)} \cdot [2/(1+n)]^{2/(1+n)} \tag{9.39}$$

C_1^* differs from C in Eq. (9.23) by a factor $[2/(1+n)]^{2/(1+n)}$. Simultaneously, the expression for the cyclic stress range corresponding to the endurance limit is also different

$$\Delta\sigma_c* = (E\sigma_f\varepsilon_f)^{0.5} \cdot (\Delta\varepsilon_c/2\varepsilon_f)^{(1+n)/2} \cdot [2/(1+n)]^{0.5} \tag{9.40}$$

Note that the difference of $\Delta\sigma_c^*$ in Eq. (9.40) from that in Eq. (9.24) is a factor of $[2/(1+n)]^{0.5}$. The fatigue strength coefficient C_1^* and equivalent stress range $\Delta\sigma_c^*$ now depend on the uniaxial tensile data of the material, and the determination of which was discussed in practical examples [17].

Similarly, in considering mean stress effect, by using SWT method and Eqs. (9.7) and (9.36) can be replaced by

$$\Delta\sigma^* = K_t\Delta\sigma_0/[2(1-R)]^{1/2}, \Delta\sigma_c^*$$
$$= [2E\sigma_f\varepsilon_f/(1-R)]^{1/2} \cdot (\Delta\epsilon_c/2\epsilon_f)^{(1+n)/2} \cdot [2/(1+n)]^{0.5} \tag{9.41}$$

Thus, Eqs. (9.37) *through* (9.41) *could be used to assess the fatigue life with mean stress effect*.

Complex forms of Eqs. (9.37) through (9.41) have been applied to analyze the fatigue behavior of several metals [17–21], including LY12CZ aluminum, 16 Mn steel and copper alloys.

9.3 Effect of Pre-strain on Fatigue

Oil and gas pipelines may contain mechanical defects during installation, manufacturing, or repair. These defects can affect the safety of the pipeline and reduce its service life, even leading to huge economic costs and endangering the surrounding ecological environment. Mechanical defects can cause local stress concentrations, which can cause local stresses to exceed plastic yielding strength and reduce load capacity. In addition, defect may affect fatigue resistance and result in early fatigue failure [4–6].

Recently, research on dents has focused on residual strength assessment and strength assessment of actual engineering. All of the above approaches are based primarily on full-scale testing.

However, research on the effects of defects on the fatigue life of pipes is limited.

Since a dented pipe is originally in general a crack-free component, its total fatigue life could be separated into two parts, namely the initiation life of fatigue crack together with the propagation life of fatigue crack.

In the previous section, a local strain fracture model for predicting fatigue crack initiation (FCI) life was proposed, and the fatigue limit was considered. Similar to the macroscopic plastic yield of a metal, it assumes the critical strain or stress level of the cyclic load, that is, the fatigue endurance limit strain exists. If the cyclic strain/deformation is lower than the critical strain, it is assumed that the component does not induce damage, and therefore, the fatigue life is infinite, or longer than 10^7–10^8, as defined.

9.3.1 Pre-strain Effect on the Initiation Life of Fatigue Crack for Pipeline Steel X60

Cosham et al. studied the effects of gouge and dent on burst strength, fatigue life, and toughness, etc. [22]. The dent was cataloged into six types, and each exhibits its special function on above material properties.

As to the pre-strain effect on the fatigue crack initiation life, pipeline steel X60 was studied [23]. The size of gauge part of the specimen is 140 mm $\times$ 28 mm $\times$ 4.8 mm, and the relative elongation for the pre-strain was fixed at $\delta_t = 5$, 10, and 15% [23], respectively.

The stress–strain curve of the original pipeline steel X60 has a plastic yielding plateau, and the ductility and reduction in cross-sectional area are about $\delta = 20.06\%$ and $\psi = 65.2\%$, respectively [23]. However, for pre-deformed steel X60, there is almost no plastic yielding plateau on the stress–strain curves [23].

Table 9.1 provides the tensile data of the pipeline steel X60 in its original and pre-strained status, which shows increases in ultimate tensile strength and plastic yielding strength of the pre-strained pipeline steel X60, while decreases the cross-sectional reduction and ductility.

The fatigue test was carried out using a sample with a single edge notch [23]. The load ratio of the minimum stress to the maximum stress in the fatigue test is fixed at $R = 0.1$ and 0.5. The fatigue load is applied with the frequency of 63 Hz. The fatigue crack initiation life was defined by the fatigue cyclic number which corresponds to a crack length of 0.25 mm appearing at the root of the notch. The variation of experimental fatigue data for C with pre-strain (5%, 10%, and 15%) is

Table 9.1 Tensile property of steel X60 in pre-deformed and original status

Pre-deformation, δ_t (%)	0	5	10	15
Elastic modulus, E (GPa)	198	198	198	198
Yield strength, σ_S (MPa)	470	543	586	601
Ultimate tensile strength, σ_b (MPa)	544	572	609	618
Relative elongation at fracture, δ (%)	20.06	17.95	12.56	8.87
Cross-sectional reduction, ψ (%)	65.20	62.63	55.87	49.02
Fracture strength, σ_f (MPa)	898	930	949	921
Fracture ductility, ε_f	1.0555	0.9843	0.8180	0.6763
Hardening coefficient, K (MPa)	894	931	960	942
Hardening exponent, n	0.077	0.062	0.059	0.056

shown in Fig. 9.5. Table 9.2 shows a comparison of the predicted fatigue resistance C and the measured fatigue resistance C for the steel X60.

The fatigue crack initiation resistance coefficient C controls fatigue crack initiation life. In the improved fatigue crack initiation life assessment, the expression of the resistance factor C is Eq. (9.39), and the comparison of the fitted test data with the predicted values of parameter C shown in Table 9.2 and Fig. 9.5 for the steel X60 indicates the validity of Eq. (9.39). The basic data for the calculation of Eq. (9.39) is taken from Table 9.1.

The tendency of C_p and C_t in Table 9.2 shows that Eq. (9.39) could show the effect of pre-strained on the fatigue crack initiation resistance coefficient of the pipeline steel. The fatigue resistant factor C depends on the pre-deformation significantly, see Fig. 9.5. The fatigue resistant factor increases with pre-deformation until the maximum uniform strain of the metal, $\delta_b = 12.6\%$; then, the fatigue resistant factor decreases with pre-deformation.

In fact, Eq. (9.39) indicates clearly that C depends on E, σ_f, ε_f, and n, while σ_f, ε_f, and n depend on pre-deformation inevitably, so the resistant factor C has a dependence on pre-deformation. Table 9.1 shows that the change in fracture

Fig. 9.5 Effect of pre-strain on fatigue crack initiation resistance coefficient of steel X60, Δ, experimental data; dashed line, prediction

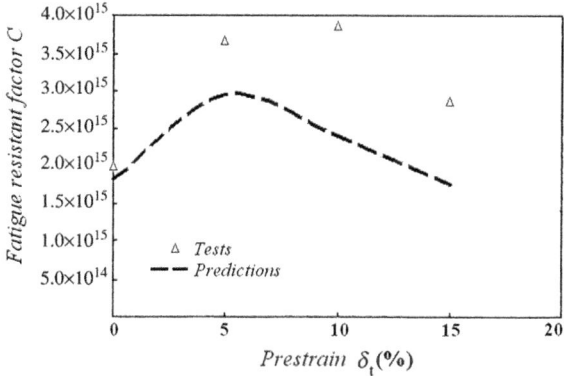

Table 9.2 Comparison of predicted and tested fatigue resistance C for steel X60

δ_t (%)	0	5	10	15
C_p	1.82×10^{15}	3.05×10^{15}	2.48×10^{15}	1.95×10^{15}
C_t	1.96×10^{15}	3.66×10^{15}	3.85×10^{15}	2.87×10^{15}

Note C_p and C_t represent the prediction and the fitted test data, stress in MPa

strength σ_f with respect to pre-strain is non-monotonic, although the fracture ductility ε_f decreases with pre-strain monotonically, and the results in Table 9.1 fairly display these phenomena.

The threshold for crack initiation $\Delta\sigma_c$ is another important parameter. The fatigue crack initiation life is infinite when the loading range is lower than $\Delta\sigma_c$.

The threshold of crack initiation $\Delta\sigma_c$ for the original pipeline steel X60 is $\Delta\sigma_c = 441$ MPa; the experimental data indicates that pre-strain induces an increase as compared to 441 MPa, see Eq. (9.42).

$$\Delta\sigma_c = 441 + 292\,\delta_t \tag{9.42}$$

Figure 9.6 shows the change in the fitted fatigue crack initiation threshold versus the pre-strain for the steel X60. The dependence of the crack initiation threshold on the pre-strain indicates the comprehensive effect of material strength and ductility [23]. As can be seen from the previous equation, $\Delta\sigma_c$ depends on σ_{-1}, E, and ε_f. On the one hand, the ultimate tensile strength of the material increases with pre-strain, the so-called metal work hardening phenomenon, and on the other hand, the ductility of the material decreases simultaneously. As a rule of thumb, for steels in the range $\sigma_b < 1800$ MPa [10, 24], there is an approximate correlation between σ_{-1} and σ_b, i.e., $\sigma_{-1} \approx 0.5\sigma_b$, and according to Eq. (9.14) and the experimental data of σ_b and ε_f in Table 9.1, it can be seen that $\Delta\sigma_c$ increases with pre-strain. This situation is clearly reflected in (9.39) and Fig. 9.6.

Fig. 9.6 Fatigue crack initiation threshold versus pre-strain for steel X60

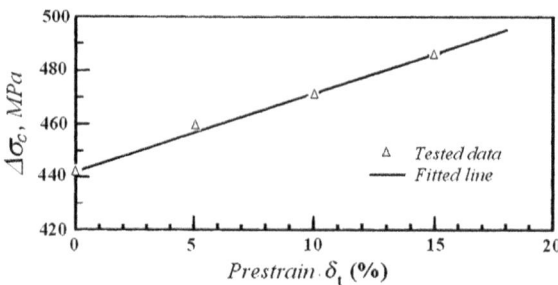

9.3.2 Summary

From the above discussion, we can draw the following conclusions:

(1) Fatigue crack initiation life depends on the strength and ductility of the metal. Pre-strain does not always reduce the fatigue life of steel; it depends on the actual content of the pre-strain.
(2) Pre-strain causes strain hardening of pipeline steel X60, increase in yield strength, and reduction in ductility.
(3) Fatigue crack initiation resistance coefficient C is influenced by pre-strain significantly. For pipeline steel X60, the fatigue resistance coefficient increases till the maximum uniform strain, $\delta_b = 12.6\%$.
(4) The threshold $\Delta\sigma_c$ of fatigue crack initiation increases with pre-deformation due to strain strengthening effect due to pre-strain.

References

1. Sih GC Jr, Moyer ET (1983) Path dependent nature of fatigue crack growth. Eng Fract Mech 17:269–280
2. Sih GC, Jeong DY (1990) Fatigue load sequence effect ranked by critical available energy. Theor Appl Fract Mech 14:141–151
3. Zheng M, Hu C, Luo ZJ, Zheng X (1994) Damage characterization of SAE 1020 and 1045 steel under torsion and compression. Theor Appl Fract Mech 21:91–99
4. Lu B, Zheng X (1992) Predicting fatigue crack initiation life of aluminium alloy at low temperature. Fatigue Fract Eng Mater Struct 15:1213–1222
5. Lu B, Zheng X (1992) An approach for predicting fatigue crack initiation life of aluminium alloy at low temperature. Fatigue Fract Eng Mater Struct 15:339–346
6. Zheng X, Lu B (1987) On the fatigue formula under stress cycling. Int J Fatigue 9:169–174
7. Zheng X (1986) A further study on fatigue crack initiation life. Int J Fatigue 8:17–22
8. Zheng X, Ling C (1988) On the expression of fatigue crack initiation considering the factor of overloading. Eng Fract Mech 31:959–966
9. Zheng X (1982) Local strain range and fatigue crack initiation life. In: IABSE Proceedings of fatigue colloquium, Lausanne, pp 169–178
10. Zheng X, Wang H, Yan J, Yi X (2013) Fatigue theory of materials and its application in engineering. Science Press, Beijing
11. Dowling NE (2013) Mechanical behavior of materials: engineering methods for deformation, fracture, and fatigue, 4th edn. Pearson Education Limited, Harlow, UK
12. Ellyin F (1985) Effect of tensile-mean-strain on plastic strain energy and cyclic response. J Eng Mater Tech 107:119–125
13. Smith KN, Watson P, Topper TH (1970) A stress-strain function for the fatigue of materials. J Mater 5:767–778
14. Dowling NE (2009) Mean stress effects in strain–life fatigue. Fatigue Fract Eng Mater Struct 32:1004–1019
15. Molski K, Glinka G (1981) A method of elastic-plastic stress and strain calculation at a notch root. Mater Sci Eng 50:93–100
16. Feltner CE, Morrow JD (1961) Microplastic stain hysteresis energy as a criterion for fatigue fracture. J Basic Eng 83:15–22

17. Zheng M, Niemi E, Zheng X (1997) An energetic approach to predict fatigue crack initiation life of LY 12CZ aluminum and 16 Mn steel. Theor Appl Fract Mech 26:23–28
18. Zheng M, Tong MX, Cai HP, Xu CZ, Huang MQ (2010) Fatigue crack initiation life of fine grain brass H62. Theor Appl Fract Mech 54:105–109
19. Wang QJ, Xu CZ, Zheng MS, Zhu JW, Du ZZ (2008) Fatigue crack initiation life prediction of ultra-fine grain chromium–bronze prepared by equal-channel angular pressing. Mater Sci Eng A 496:434–438
20. Xu CZ, Wang QJ, Zheng MS, Zhu JW, Li JD, Huang MQ, Jia QM, Du ZZ (2007) Microstructure and properties of ultra-fine grain Cu–Cr alloy prepared by equal-channel angular pressing. Mater Sci Eng A 459:303–308
21. Wang QJ, Xu CZ, Zheng MS, Zhu JW, Buksa M, Kunz L (2007) Fatigue property of ultrafine-grained copper produced by ECAP. Acta Metall Sin 43(5):498–502
22. Cosham A, Hopkins P (2004) The effect o dents in pipelines—guidance in the pipeline defect assessment manual. Int J Press Vessel Pip 81:127–139
23. Zheng M, Luo JH, Zhao XW, Bai ZQ, Wang R (2005) Effect of pre-deformation on the fatigue crack initiation life of X60 pipeline steel. Int J Press Vessel Pip 82:546–552
24. Zheng X (2001) On some basic problems of fatigue research in engineering. Int J Fatigue 23:751–766

Chapter 10
Energy Absorption of Highly Ductile Materials

Abstract The energy absorption of highly ductile materials and characteristics of energy-absorbing components are presented first, and then both horizontally compressed ring and axial compression of round tube are given especially.

10.1 Introduction

Energy absorption refers to the process of dissipating impact energy in the form of deformation and damage such as plastic deformation or brittle fracture in an energy-absorbing element or structure during a collision event. The energy absorption behavior of materials and structures plays a key role in the safety of the structure after impact. In practical engineering equipment such as aerospace, automobiles, rail vehicles, offshore platforms, highway guardrails, and nuclear power plants for the purpose of safety protection, the energy absorption performance of the structure must meet strict requirements. This chapter focuses on energy-absorbing components made of ductile materials (such as low carbon steel, aluminum, and polymers) and introduces the application of plastic deformation principles in energy absorption design.

A lot of work has been done to study the loading ability and energy adsorption ability of tubes these years. Such absorbers can be used in the form of circular tubes, square tubes, frusta, struts, honeycombs, and sandwich plates. Circular or square sectional tubes are the most common structure of the energy absorber due to its easy manufacturability. Tubes can exhaust both elastic and inelastic energy through different modes of deformation, which results in different energy absorption responses.

The deformation involved in energy absorption processes includes axial crushing and lateral compression (flattening) mainly. The actual application of this type of energy absorbers covers automobile, nuclear reactors, aircraft, spacecraft industries, etc. [1].

Niknejad et al. gave a theoretical relation to predict the instantaneous folding force in the hexagonal columns under axial loading [2–4], using basic folding

© Springer Nature Singapore Pte Ltd. 2019
M. Zheng et al., *Elastoplastic Behavior of Highly Ductile Materials*,
https://doi.org/10.1007/978-981-15-0906-3_10

mechanism. It indicated that the folding process in the rectangular and hexagonal column is the best choice in the type of hollow structural energy absorbers. Furthermore, Niknejad et al. studied the effect of foam filled on the lateral compression in the circular tubes [5].

Reid and Reddy proposed a theoretical relation to predict the effect of strain hardening on the lateral compression and plastic bending of circular tubes during the large deformations with free sides [6, 7]. Avalle and Goglio applied strain gauges on circular tubes to represent the experimental insight regarding the strain field during lateral compression and check the previous theoretical models [8]. Leu employed an incremental elastoplastic finite element method to examine the lateral compression of aluminum and clad tubes [9]. The effects of various process parameters on the occurrence of buckling of tube were studied [9]. Nemat-Alla proposed a simple technique to reflect the stress–strain behavior of laterally compressive tube [10]. Gupta et al. presented detailed experimental and computational investigations of metallic round tubes under quasi-static loading [11]. Morris et al. investigated lateral compression in nested circular and elliptical energy absorbers experimentally and numerically [12]. Celentano and Chaboche analyzed the damage evolution in a steel cylinder during the flattening process [13]. The crushing behavior and energy absorption capabilities for mild steel nested tubes were experimentally analyzed and numerically simulated [14, 15]. Zeinoddini et al. studied the behavior of axially preloaded steel tubes under condition of lateral impacts [16]. Zuraida et al. analyzed the composite tubes under quasi-static lateral indentation loading experimentally and numerically [17]. Mcdevitti et al. analyzed the elastic–plastic ring between rigid plates by linear hardening material model [18], and complicated expressions were obtained. Lee et al. conducted experiments to study the energy absorption characteristics of thin-walled square tubes under the condition of quasi-static axial loading in order to develop the optimum structural members, a controller was introduced to improve and control the absorbed energy of thin-walled square tubes [19].

Recently, the single-tube and multi-member tubular structures subjected to lateral compressive experiments and FE simulation were conducted by Lipa and Kotełko [20]. FE modeling was used to simulate the deformation behavior and energy absorption of circular tube under lateral loading [21, 22]. Energy absorption characteristics of foam-filled multi-cell thin-walled structure (FMTS) with nonlinear FE analysis were investigated by Yin et al. [23]. Zahiri-Hashemi et al. studied the seismic lateral load distribution pattern of 3-, 6-, 12-, and 16-story buckling restrained steel braced frames [24]. It concluded that the code pattern lateral load is quite different from that of analytical lateral load pattern in particular for taller structures except for low-rise structures. Two different versions of the well-known 2-DOF Augusti model were studied by Gantes et al. analytically [25]. FE calculation of energy absorbing in ANSYS/LS-DYNA for the structures of three types of aluminum honeycomb (honeycombs 1, 2, and 3, respectively) was conducted by Xie and Zhou numerically [26]. It found that larger the plateau stress acting on the honeycomb leads to greater contribution of the honeycomb in overall energy dissipation of the structure.

In general, the design and analysis of energy-absorbing structures are completely different from the design and analysis of conventional structures in order to dissipate impact kinetic energy in a controlled manner or at a predetermined rate. In order to achieve this, the following basic principles should be followed in the design [27]:

(1) The energy conversion achieved by structural and material deformation should be irreversible, i.e., the structure and material should be capable of converting most of the input kinetic energy into inelastic energy through plastic deformation or other dissipative processes. In other words, the energy-absorbing element must dissipate the kinetic energy to achieve the intended purpose. Instead of being elastically compressed like a spring, it absorbs the ability first and then bounces back to release the absorbed capacity. The effect of the energy-absorbing element is completely opposite to the action of the spring.

(2) In the process of large deformation, it should not only have sufficient energy absorption capacity, but the peak value of reaction force of the energy absorption structure or material in the collision should be lower than the damage threshold value of the protected objects; it is better to keep the reaction force be constant or almost constant to avoid excessive deceleration rates and so as to maintain the structural and personnel safety.

(3) There is a great deal of uncertainty due to the magnitude, pulse form, direction, and distribution of the external loads acting on the energy-absorbing structure and material. Thus, these structures and materials should have a stable, repeatable deformation pattern to ensure structural reliability under complex operating conditions.

(4) The energy-absorbing element itself should be lightweight and have a high specific energy absorption capacity, which is extremely important for the carrying of vehicles. For various personal protective devices (such as helmets and body armor), lightweight is also very important.

(5) Energy-absorbing devices are usually disposable devices, that is, once they are deformed after use, they are discarded and replaced. Therefore, the manufacture, installation, and maintenance of these energy-absorbing devices should be easy and cost effective.

In view of the above requirements of energy-absorbing systems, their mechanical behavior must be characterized by plastic deformation of materials and structures. Complex absorption systems are primarily composed of elements with energy-absorbing properties. In addition to the commonly used rings and tubes, some common energy-absorbing elements and structures are combined. In this chapter, we will present the application of plastic deformation principle in the design of energy-absorbing components by taking circular tubes and square tubes as examples.

10.2 Horizontally Compressed Rings and Tubes

10.2.1 Ring Pressed Between Two Rigid Plates

As early as 1963, de Runtz and Hodge discussed the large deformation of the ideal rigid plastic ring pressed between two rigid plates. They got the plastic deformation mechanism of the ring under the condition of neglecting axial force, shear force, elasticity, strengthening, and friction, which can be described by a four-hinge model, as shown in Fig. 10.1a. For the balance of a circular arc (see Fig. 10.1b), it can be concluded that [27]

$$F = F_c/\cos\theta, \delta = 2R\sin\theta. \tag{10.1}$$

in which $F_c = 4Mp/R$ is the initial ultimate load of the ring, F is the load in the case of large deformation, δ is the relative displacement of the two rigid plates, and R and D represent the radius and diameter of the ring, respectively. By eliminating δ from Eq. (10.1), the load–displacement relationship under large deformation can be obtained as

$$F = F_c/[1 - (\delta/D)^2]^{1/2} \tag{10.2}$$

or

$$F = 2\sigma_y h^2 L/\left\{D[1 - (\delta/D)^2]^{1/2}\right\} \tag{10.3}$$

in which h is the thickness of the ring, and L is the width of the ring. Equation (10.2) shows that the load increases with increasing deflection (see

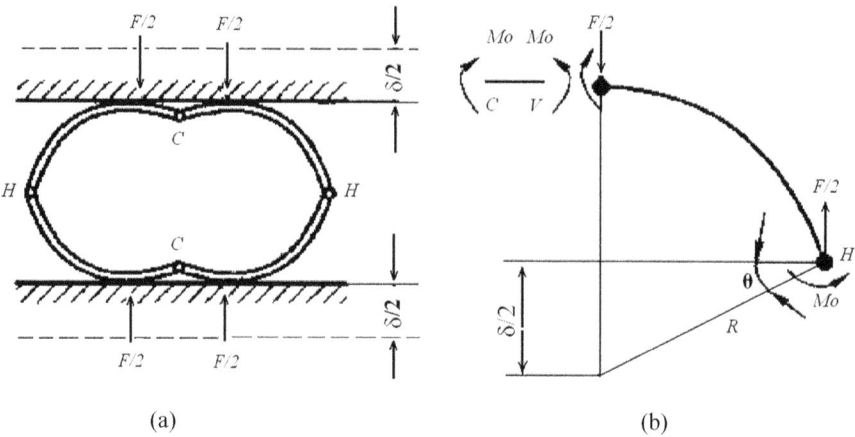

(a)   (b)

Fig. 10.1 Four-hinge mechanism of ring suffering large deformation. **a** The destructive mechanism proposed by de Runtz and Hodge. **b** The force acting on the deformed arc [27]

Fig. 10.2). It is worth to note that the above analysis of the ring is equally applicable to round tubes under similar loads, as long as the appropriate yield strength σ_y is chosen. The σ_y in Eq. (10.3) can take the yielding strength obtained by a simple tensile test. When the length of the tube is greater than its diameter, considering the plane strain condition, σ_y should be taken as yielding strength in simple tensile test multiplied by a factor $2/3^{1/2}$ [27].

After the pioneering study of de Runtz and Hodge, many people carried out experiments and found that the bearing capacity of the experimentally obtained ring after deformation was higher than that predicted by Eq. (10.3) (as shown in Fig. 10.2). This phenomenon can be attributed to strain hardening. Reid et al. have shown that the result of linear strengthening of the material not only increases the plastic limit bending moment, but also expands the plastic hinge to form a plastic zone, i.e., the hinge has a certain length. Although this length is small, important geometric changes could change the length of the arm, resulting in a change in load. A simple way to estimate strain hardening is to estimate the average strain of the deformation zone involved and then introduce it into the relationship that enhances the bending moment resistance.

The equation proposed by Redwood is

$$F/F_c = \left\{1/[1-(\delta/D)^2]^{1/2}\right\} \cdot \left[1 + E_p/[(3\sigma_y\lambda)\arcsin(\delta/D)]\right] \tag{10.4}$$

In the experiment, by measuring the plastic zone of the arc, it is found that the value of parameter λ is 5, and the corresponding curve can be drawn by using this value, as shown in Fig. 10.2.

Obviously, Eq. (10.4) is more accurate than Eq. (10.3), but it is still lower than the experimental results, especially for the case of large deflection. This is because plastic hinges are still considered to be localized. After Reid and Reddy studied this problem, the theory of plastic hinge (plastica) was proposed, the original hinge was replaced by a circular arc, and the length of the arc changed with the deflection of δ. The shape of the load–deflection curve given in the literature can be determined by the following dimensionless parameters [27]:

Fig. 10.2 Comparison of the experimental and theoretical curves of load versus displacement ($h/R = 0.108$, $R = 42.16$ mm, $L = 101.6$ mm) [27]

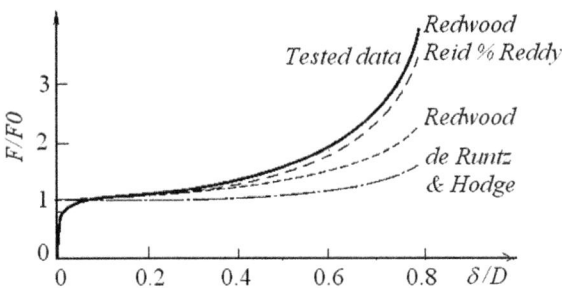

$$mR = (6\sigma_y/E_p hR)^{1/2} \tag{10.5}$$

A larger mR value corresponds to a relatively flat load–deflection curve; a smaller mR value causes the curve to rise significantly. Compared with the previous theory, their theory is in good agreement with the experimental results [27].

10.2.2 Elastic Limit Analysis of Elliptical and Circular Tubes Under Lateral Force Condition

In this section, elastic limit analysis newly developed by Zheng M. et al. is employed to analyze the bearing capacity of elliptical and circular tubes under lateral forces [28]; and the elastic limit loads of elliptical and circular tubes under lateral forces were given analytically [28]. Experimental results from the existing literature for steel tubes and aluminum tubes were cited as applications of the proposed expression [28].

Figure 10.3 shows a thin-walled elliptical tube that is subjected to symmetric lateral tension.

Due to the symmetry of the component and load, a quarter was studied only, see Fig. 10.3b, and the cross section mn is shear-free and has a tensile force $P/2$. M_0 is the bending moment on the cross section mn, but its value is undetermined.

Therefore, the cross section mn is rotation-free owing to the symmetry of the structure, and the displacement corresponding to M_0 is zero [29], i.e.,

$$dU/dM_0 = 0 \tag{10.6}$$

where U is the strain energy of a quarter elliptical tube when it is loaded.

Fig. 10.3 Symmetrical transverse tension applied to elliptical thin-walled tube

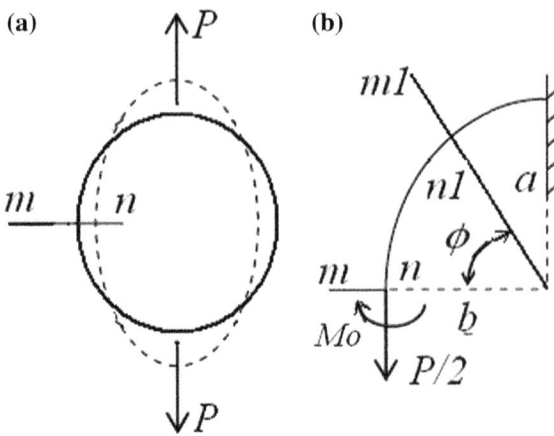

The bending moment of the cross section $m1n1$ in the direction of the angle ϕ is, see Fig. 10.3b,

$$M = M_0 - \frac{P}{2}b(1 - \cos\phi) \tag{10.7}$$

in which b is the half length of the short axis of the ellipse, and

$$dM/dM_0 = 1 \tag{10.8}$$

At the same time, the potential energy of deformation is [29]

$$U = \int_0^S \frac{M^2}{2EI_Z}ds \tag{10.9}$$

in which E is the elastic modulus of the tube, I_z is the moment of inertia, and ds is the increment of the elliptical arc.

Substituting Eqs. (10.7) through (10.9) into Eq. (10.8), it yields

$$0 = \int_0^{S_{\pi/2}} \left[M_0 - \frac{P}{2}b(1 - \cos\phi) \right] ds$$

$$= \int_0^{\pi/2} \left[M_0 - \frac{P}{2}b(1 - \cos\phi) \right] \cdot \sqrt{a^2 \cos^2\phi + b^2 \sin^2\phi} \cdot d\phi \tag{10.10}$$

in which a is the half length of the long axis of the ellipse.

Introducing the new parameters R and ζ to characterize the shape of the ellipse by the following definitions:

$$a = R(1 + \varsigma), \quad b = R(1 - \varsigma) \tag{10.11}$$

in which R represents the radius of the corresponding perfect circular tube, and ζ represents the degree of deviation from the complete circular tube.

Obviously, the shape of the ellipse varies with the parameter ζ. When the parameter ζ is 0, it emerges as a perfect circular tube of radius R. Therefore, the value of value ζ represents the ellipticity of the ellipse.

Finally, from Eq. (10.10), it derives

$$M_0 = \frac{P \cdot R \cdot (1 - \xi)}{2} \cdot \left\{ 1 - \frac{1}{\pi(1 + \varsigma^2/4)} \cdot \left[1 - \varsigma + \frac{(1 + \varsigma)^2}{\sqrt{4\varsigma}} \cdot \arcsin\left(\frac{\sqrt{4\varsigma}}{1 + \varsigma}\right) \right] \right\} \tag{10.12}$$

Substituting Eq. (10.12) into Eq. (10.7), it obtains the bending moment on the cross section $m1n1$ from the angle of ϕ, see Fig. 10.3b),

$$M = M_0 - \frac{P \cdot b}{2}(1 - \cos \varphi)$$

$$= \frac{P \cdot R \cdot (1 - \varsigma)}{2} \cdot \left\{ \cos \varphi - \frac{1}{\pi(1 + \varsigma^2/4)} \cdot \left[1 - \varsigma + \frac{(1 + \varsigma)^2}{\sqrt{4\varsigma}} \cdot \arcsin\left(\frac{\sqrt{4\varsigma}}{1 + \varsigma}\right) \right] \right\}$$

$$(10.13)$$

For the loading points, $\phi = \pi/2$ in Fig. 10.3a), it derives the maximum bending moment for a given loading P, i.e.,

$$M_{\pi/2} = -\frac{P \cdot R \cdot (1 - \varsigma)}{2} \cdot \frac{1}{\pi(1 + \varsigma^2/4)} \cdot \left[1 - \varsigma + \frac{(1 + \varsigma)^2}{\sqrt{4\varsigma}} \cdot \arcsin\left(\frac{\sqrt{4\varsigma}}{1 + \varsigma}\right) \right]$$

$$(10.14)$$

In turn, when the load P reaches to its critical value P_e, plastic yielding occurs at the loading point $\phi = \pi/2$ first in Fig. 10.3a

$$P_e = \frac{2M_e}{R} \frac{\pi(1 + \varsigma^2/4)}{\left[1 - \varsigma + \frac{(1 + \varsigma)^2}{\sqrt{4\varsigma}} \cdot \arcsin\left(\frac{\sqrt{4\varsigma}}{1 + \varsigma}\right) \right] \cdot (1 - \varsigma)}$$

$$(10.15)$$

in which M_e is the plastic bending moment of a sheet with the thickness t and width l [28]

$$M_e = \frac{\sigma_s t^2 l}{4}$$

$$(10.16)$$

where σ_s is the plastic yielding strength of the tube material. Here, elastic–perfect plastic material model is employed.

Substituting Eq. (10.11) into Eq. (10.15), it obtains

$$P_e = \frac{\sigma_s t^2 l}{2R} \frac{\pi(1 + \varsigma^2/4)}{\left[1 - \varsigma + \frac{(1 + \varsigma)^2}{\sqrt{4\varsigma}} \cdot \arcsin\left(\frac{\sqrt{4\varsigma}}{1 + \varsigma}\right) \right] \cdot (1 - \varsigma)}$$

$$(10.17)$$

Equation (10.17) is the critical load estimation for elliptical tubes under lateral tension, and it indicates that the value of P_e increases with the increase of ζ.

From Eq. (10.17), for a perfect circular tube, $\zeta = 0$, thus it obtains the critical load for a circular tube under lateral tension

$$P_{Ce} = \frac{\pi \sigma_s t^2 l}{4R} \tag{10.18}$$

It can be seen from Eq. (10.18) that the critical load value of the perfect circular tube under transverse tension is not greater than the corresponding elliptical tube under lateral tension. An alternative form of Eq. (10.17) is

$$P_e / [\sigma_s(t \cdot l)] = \frac{\pi t}{2R} \frac{(1 + \varsigma^2/4)}{\left[1 - \varsigma + \frac{(1+\varsigma)^2}{\sqrt{4\varsigma}} \cdot \arcsin\left(\frac{\sqrt{4\varsigma}}{1+\varsigma}\right)\right] \cdot (1 - \varsigma)} \tag{10.19}$$

The term $t \cdot l$ in the Eq. (10.19) represents the cross-sectional area of the tube sheet having a thickness t and a width l. Therefore, the term $P_e / [\sigma_s (t \cdot l)]$ in Eq. (10.19) gives the normalized critical load of the elliptical tube under transverse tension, which increases with the increase of the parameters t/R and ς definitely. Figure 10.4 shows the variation of the normalized critical load versus the parameters t/R and ς.

10.2.3   Experimental Test of Critical Load Estimation for Elliptical Tube Under Lateral Load

If a circular or elliptical tube is subjected to a symmetrical tensile load, plastic yielding occurs at the loading point $\phi = \pi/2$ first (Fig. 10.3b), and it becomes a new elliptical tube with a larger ς, i.e., the shape parameter ς gets a small increment $\Delta\varsigma$.

Since the critical load P_e of the elliptical tube increases with ς under lateral tension, Eq. (10.17) can be used to estimate the limit or critical load of a circular or elliptical tube subjected to a tensile load, which may reflect an elliptical forming of the tube progressively from a circular or elliptical tube.

In fact, if we ignore all minor effects, such as tube length, friction, and heating during the experiment, the actual critical load may be the same for tubes with the same instantaneous shape, and it does not depend on whether the loading type is pulled or compressed one.

Fig. 10.4 Variation of the normalized critical load with respect to t/R and ς

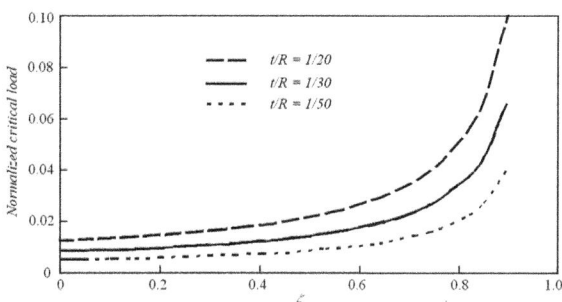

Gupta et al. represented the experimental and computational study of quasi-static lateral compression of circular metal tubes [11] and discussed the influence of process parameters on the deformation behavior of the tube.

Experimental data from Gupta et al. was used to check the estimation formula of the critical load of the elliptical tube under lateral compression [28].

Table 10.1 shows the tubular parameters used in Gupta experiments and calculations [11].

Substituting the material parameters from Table 10.1 into Eq. (10.17), which produces the critical loads for the S504 and A503 tubes under lateral compression, Fig. 10.5 gives the comparison.

Similarly, Lipa and Kotełko performed a lateral compression experiment on a single tube, which was employed to check the expression again [20]. Table 10.2 shows the tubular parameters used in the Lipa and Kotelkko's tubular experiments [20]. Figure 10.6 shows a comparison of the predictions of the Eq. (10.17) with the experimental results of Lipa and Kotełko; the theoretical results of de Runtz and Hodge are given also.

Figures 10.5 and 10.6 show that the predictions from Eq. (10.17) agree with the experimental data fundamentally.

10.3   Axial Compression of Round Tubes

10.3.1   Axisymmetric Large Deformation of a Circular Tube—the Alexander Model of the Ring Pattern

In the analysis of the bearing capacity of the ring mode, it assumed that the material is ideally rigid plastic, and the process of the axisymmetric flexion of the round tube can be simplified into the model shown in Fig. 10.7 [27]. At this time, the energy dissipated can be the superposition of the following two parts.

The bending strain energy is

$$W_b = 2M_0 \cdot \pi D \cdot \frac{\pi}{2} + 2M_0 \int_0^{\pi/2} \pi(D + 2H \sin \theta)d\theta = 2\pi M_0(\pi D + 2H) \quad (10.20)$$

Table 10.1 Material parameters of laterally compressive round tubes S504 and A503 [11]

Sp. No.	S504	A503
Average radius (R) (mm)	22.4	23.83
Thickness (t) (mm)	3.34	3.44
R/t	7.120	6.925
Width (l) (mm)	100	100
Material properties σ_s (MPa)	288.2	153.3

Fig. 10.5 Comparison of the estimation and laterally compressive tests for circular tubes S504 and A503

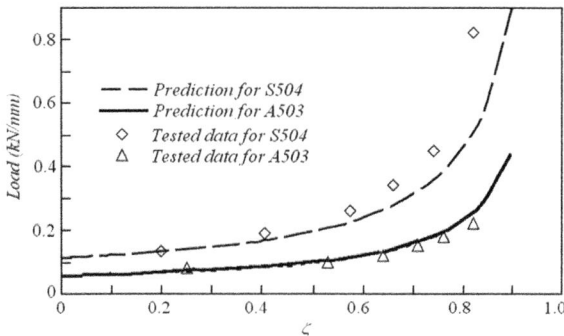

Table 10.2 Tubular parameters of Lipa and Kotełko's experiments [20]

Thickness t (mm)	Material properties σ_s (MPa)	Tube diameter 2R (mm)	Width l (mm)	R/t
3.2	378	88.9	100	13.891

Fig. 10.6 Comparison of predictions of de Runtz and Hodge and Eq. (10.17) with Lipa and Kotełko's test results

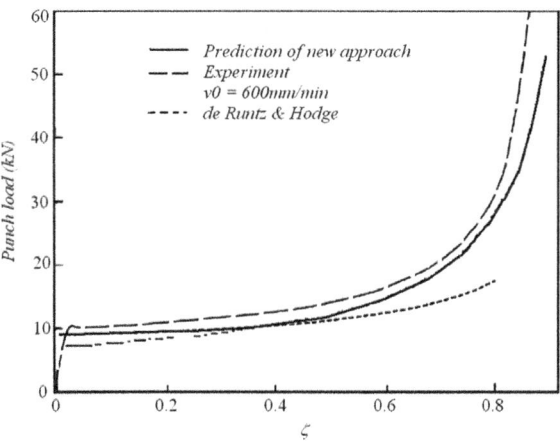

Fig. 10.7 Axisymmetric flexural model

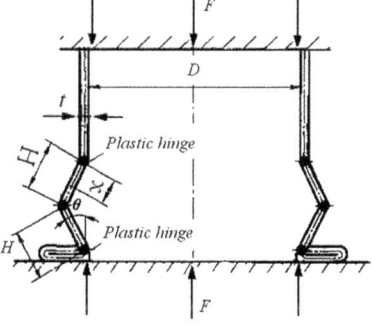

in which H is the bend length to be determined, and M_0 is the plastic limit bending moment of the unit width of the pipe wall. According to the Mises yield condition, $M_0 = \frac{2}{\sqrt{3}} \cdot \frac{1}{4} \sigma_s t^2$, where t is the wall thickness, and σ_s is the yield strength.

At the same time, the tensile strain deformation energy in the middle section of the tube is

$$W_a = 2 \int_0^H \sigma_s \cdot \pi D t dx \cdot \ln[(D + 2x \sin \theta)/D], \tag{10.21}$$

when the folding angle is 90°, it is obtained from Eq. (10.21)

$$W_a = 2\pi \sigma_s t H^2. \tag{10.22}$$

In the above derivation, it assumes that there is no interaction between the bending moment and the membrane force, and that both c and H remain unchanged during the deformation process. At this time, consider the energy balance of the process of generating a flexion

$$F_m \cdot 2H = W_b + W_a, \tag{10.23}$$

in which F_m is the average of the loads during the generation of a flexion. Substitute Eqs. (10.21) and (10.22) into the Eq. (10.23), it yields

$$F_m / \sigma_s = \frac{\pi t^2}{\sqrt{3}} \left(\frac{\pi D}{2H} + 1 \right) + \pi H t. \tag{10.24}$$

According to the idea that the real H should make F_m minimum, H can be determined by $\partial (F_m/\sigma_s)/\partial H = 0$. Therefore,

$$H = \left(\frac{\pi}{2\sqrt{3}} \right)^{1/2} (Dt)^{1/2} = 0.95(Dt)^{0.5}. \tag{10.25}$$

Substitute Eq. (10.25) into Eq. (10.24), an average load is obtained

$$F_m / \sigma_s = 6t(Dt)^{0.5} + 1.8t^2. \tag{10.26}$$

The above formula is obtained by assuming that the material of the bending tube is toward the outside of the tube after buckling. Similar results can be obtained if the material is assumed to be bent into the inside of tube after buckling

$$F_m / \sigma_s = 6t(Dt)^{0.5} - 1.8t^2. \tag{10.27}$$

The actual situation is usually between these two extreme cases, so it is reasonable to approximate the average of the carrying capacity as

$$F_m/\sigma_s = 6t(\text{Dt})^{0.5}. \tag{10.28}$$

This is what Alexander proposed in 1960 for a symmetric crushing analysis model along the axis of the tube [30]. This model is very simple, but it captures the main phenomena observed in the experiment.

10.3.2   Non-Axisymmetric Large Deformation Crushing Theory of Circular Tubes

As early as the study of the elastic stability of the cylindrical shell under the axial pressure, it is noted that the non-axisymmetric buckling mode of the waveform is in the axial and circumferential directions simultaneously. Further research has shown that this mode can also be produced under plastic buckling conditions and develops into a fold that is folded by several triangular blocks under large deformation. Therefore, this buckling mode is also called the diamond mode [27].

The theoretical model of the diamond model is not as successful as the ring model. Johnson, Soden, and AI. Hassani developed a diamond model theory based on the experiment of PVC round tube. From the pure geometric relationship, the regularity of the triangular plastic hinge line can be determined. For example, Fig. 10.8 shows the distribution of the dump line with the toroidal fold. After calculating the energy dissipated along these plastic hinge lines, they proved that there is

$$F_m/2\pi M_0 = 1 + n\csc(\pi/2n) + n\cot(\pi/2n) \tag{10.29}$$

in which n is the number of petals that are bent in the direction of the ring. $M_0 = \frac{2}{\sqrt{3}}\frac{1}{4}\sigma_s t^2$ is the plastic limit bending moment of Mises yield condition, and σ_s is the yield stress of the material during uniaxial stretching. Although Eq. (10.29) shows the dependence of the average cutting on the number n of the flexion flaps, it is not ideally compatible with the experimental results since this analysis does not consider the actual mid-surface tensile strain energy. It is not possible to determine the dependence of the number of petals on the geometry of the tube based on this analysis.

Fig. 10.8 Non-axisymmetric modes ($n = 3$)

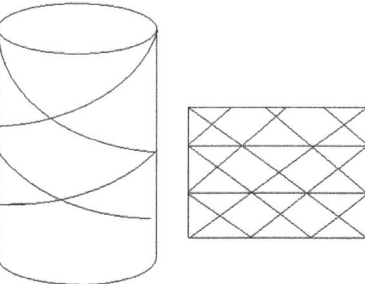

10.4   Analysis of Types and Characteristics of Energy-Absorbing Components

The energy absorption behavior of materials and structures plays a pivoting role in the safety of the structure after impact [27]. The choice of absorbent material and the design of the energy-absorbing structure will vary with the operating conditions, and there is no energy absorption structure that is suitable or absolutely optimal everywhere. The following is a brief comparison of various types of energy-absorbing elements. Figure 10.9 compares the specific energy consumption. Figure 10.10 compares the relative strokes. The specific energy consumption refers to the energy value of the energy dissipation per weight; the relative stroke is the ratio of the effective stroke to the length of the component. These two indicators are important for practical applications. Of course, there are other indicators, such as the stability of the carrying capacity mentioned above.

The effective design of energy-absorbing structure by the principle of plastic deformation is a new field of plastic mechanical application in engineering, and it has been widely developed in recent years. For example, the energy-absorbing structure is used to improve the impacting resistance performance of the vehicle; for the safety protection of highways; for the protection of industrial accidents; for personal safety protection and the like.

10.5   Effect of Loading Speed on Structural Energy Absorption Performance

10.5.1   Effect of Loading Speed on Deformation Mode—Ring In-Line System with End Impact

The quasi-static response of the ring and the tube is described above. Under the action of the dynamic load, the response of the structure is quite different [27].

Fig. 10.9 Specific energy consumption of various components

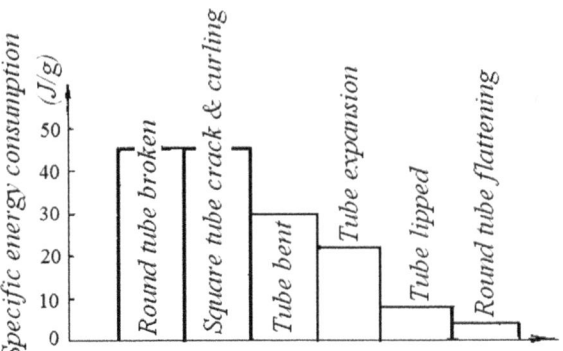

Fig. 10.10 Relative stroke
length of various components

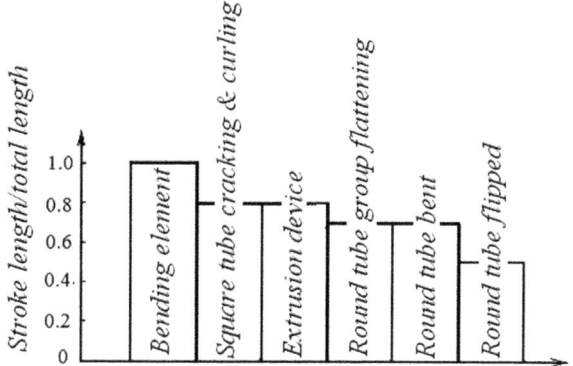

When the speed is very low, the deformation of all the rings occurs uniformly at the
same time, so the overall response of the system can be integrated by the analysis of
each ring under quasi-static. Silva-Gomes et al. studied the metal ring chain with an
impact on the end. The typical mass ratio is $m/G = 0.01$, and the velocity is $v_0 =$
4 m/s. Here, m is the mass of each ring, and G is the mass of the impactor. Reid
et al. conducted a detailed research on the collision response of the brass ring
system at higher speeds (30–120 m/s). Zhao et al. also studied the dynamic
behavior of the 6060T5 aluminum ring in-line system under impact load. It shows
that the deformation of the ring is not uniform. It starts from the four-hinge
deformation mode of the first ring and spreads to the far end gradually.

Reid et al. suggest that this phenomenon can be explained by the theory of shock
waves. The wave is generated at the impact end, and it spreads to the far end. The
early deformation of the distal ring may be the result of superimposing the
fixed-end elastic incident wave and the reflected wave, and causing the return wave
of the plastic wave. Consider a ring system where the ends are impacted by mass
G. Let the initial crushing load of a loop be F_0. The current load is F. Assume that
the quasi-static relationship $F/F_0 = f(\delta/D)$ is still true in the dynamic case, where f
(δ/D) is a function. Ignore the elastic deformation and allow the discontinuity of the
force at the contact point between the i-th ring and the $i + 1$th ring that is being
deformed. Then, some relevant relationships are established.

For the propagation of the initial shock wave across the i-th ring, the governing
equation is

$$[G + (i - 1)m](u_i - u_{i-1}) = -F_i t_i \tag{10.30}$$

$$mu_i = (F_i - F_0)t_i \tag{10.31}$$

$$\delta_i = (u_i + u_{i-1})t_i/2 \tag{10.32}$$

$$F_i/F_0 = f(\delta/D) \tag{10.33}$$

in which t_i is the time that the shock wave passes through the i-th ring; u_{i-1} and u_i are the toughness and final velocity of the system deformation part during the t_i time interval, respectively, and F_i is the force that is obtained from the load–deflection curve of the ring with the dimension being a unit. The first two equations are the conservation of momentum about the deformed part of the system and the i-th ring. Equations (10.30) through (10.33) can be solved numerically; thus, the force, velocity, and compression of the ring are obtained.

The predicted values of the ring compression by this shock wave theory are compared with the experimental results. The prediction value is about 10% higher than the experimental value [27].

10.5.2 Sensitivity of Structure Type to Impact Velocity

Under impact conditions, the metal structure absorbs energy through large plastic deformation. Its energy absorption characteristics depend on the type of structure inevitably [27]. Two types of energy absorption structures can be divided according to the shape of the load–displacement curve. The load–displacement curve of the first type of structure is with a platform in the plastic phase; see Fig. 10.11a; while the second type of structure drops sharply after the load reaches the extreme value; see Fig. 10.11b. The sensitivity of these two types of structures is different in deformation to impact velocity.

The difference between the sensitivity of the type 1 and type 2 structures with respect to the collision speed is explained below. Assume that the length dimension of the model is $1/A$ of the prototype. Due to the strain rate effect and the inertia effect, the dynamic limit load of the small size model is scaled up to be higher than the dynamic limit load of the prototype. When the collision energy remains proportional, the final displacement of the scaled model is less than the final displacement of the prototype. If the ultimate load remains constant during large deformations (this is

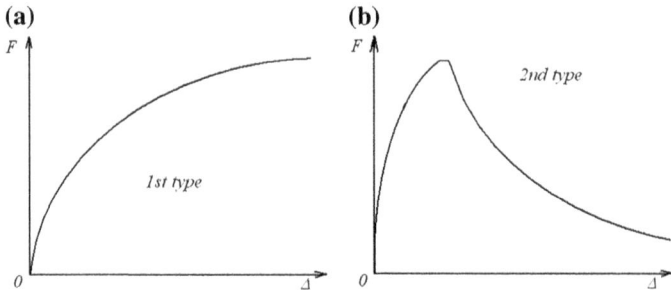

Fig. 10.11 Two types of structural load–displacement curves

typical for type 1 structures), the difference between the model and the prototype at the final displacement is not significant under equal amplification of energy [27]. However, if the load–displacement curve is in the form of "dropping sharply," then according to the same rule, this difference will become very significant. Obviously, distinguishing between these two types of energy-absorbing structures and the understanding of speed sensitivity of the second type of structure are all crucial, which is beneficial to the design of the energy absorption structure and determination of the scale law of the model test. In fact, the above discussion concerning circular tubes under lateral loading type 1 structure. Focusing on the research below is the static phase dynamic behavior of the second structure.

A simple and typical structure of type 2 is a pair of pre-bent panels. It is referred to below as a "crooked plate" to analyze its static and dynamic behavior; see Fig. 10.12 [27].

Zhang T. and Yu T. proposed a simple model that considers the energy loss during collision based on the classical theory of inelastic collision between two objects [31]. This model contains a rigid impactor of mass G and a pair of ideal rigid plastic flaps. Each plate has a length of 2 l, a mass of m, and an initial folding angle of θ_0, total deflection δ, and vertical displacement Δ; see Fig. 10.13. It is assumed that the folding plate is deformed to form a four-hinge mechanism, and the bending moments of the plastic hinge are M_1 and M_2, respectively. After lengthy derivation, they obtained

$$\frac{K_0}{T} = 1 + \frac{1}{4R_M}\left(1 + \frac{1}{3\theta_0^2}\right) \approx 1 + \frac{1}{12RM\theta_0^2} \tag{10.34}$$

in which, $K_0 = Gv_0^2/2$ is the kinetic energy carried by the impactor; T is the system kinetic energy at the moment after the collision; mass ratio $R_M = G/4$ m presents impact mass/specimen quality. Taking in mind $\theta_0 \ll 1$, it obtained Eq. (10.29). Their research got good achievements. Figure 10.13 depicts the dependence of T/K_0 on the impact velocity v_0, where $T_0 = 122$ J remains constant.

Fig. 10.12 A pair of pre-bent panels

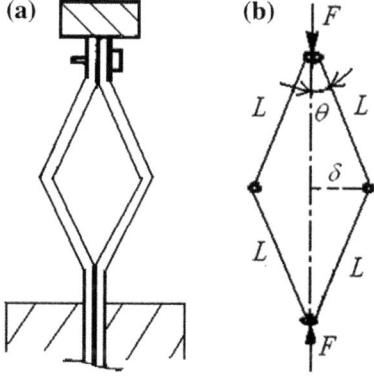

Fig. 10.13 T/K_0 versus v_0 at $K_0 = 122$ J

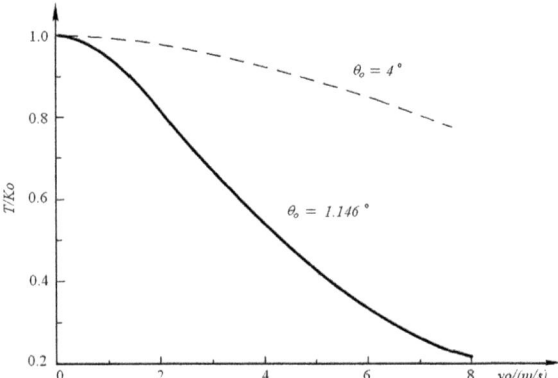

References

1. Olabi AG, Morris E, Hashmi MSJ (2007) Metallic tube type energy absorbers: a synopsis. Thin Walled Struct 45:706–726
2. Niknejad A, Liaghat GH, Moslemi Naeini H, Behravesh AH (2010) A theoretical formula for predicting the instantaneous folding force of the first fold in a single cell hexagonal honeycomb under axial loading. Proc Inst Mech Eng C J Mech Eng Sci 224, 2308–2315
3. Niknejad A, Liaghat GH, Moslemi Naeini H, Behravesh AH (2010) Experimental and theoretical investigation of the first fold creation in thin-walled columns. Acta Mech Solida Sin 23:353–360
4. Niknejad A, Liaghat GH, Moslemi Naeini H, Behravesh AH (2011) Theoretical and experimental studies of the instantaneous folding force of the polyurethane foam-filled square honeycombs. Mater Des 32:69–75
5. Niknejad A, Elahi SA, Liaghat GH (2012) Experimental investigation on the lateral compression in the foam-filled circular tubes. Mater Des 36:24–34
6. Reid SR, Reddy TY (1978) Effect of strain hardening on the lateral compression of tubes between rigid plates. Int J Solid Struct 14:213–225
7. Reddy TY, Reid SR, Carney JF, Veillette JR (1987) Crushing analysis of braced metal rings using the equivalent structure technique. Int J Mech Sci 29:655–668
8. Avalle M, Goglio L (1997) Static lateral compression of aluminum tubes: strain gauge measurements and discussion of theoretical models. J Strain Anal Eng 32:335–343
9. Leu DK (1999) Finite-element simulation of the lateral compression of aluminum tube between rigid plates. Int J Mech Sci 41:621–638
10. Nemat-Alla M (2003) Reproducing hoop stress–strain behavior for tubular material using lateral compression test. Int J Mech Sci 45:605–621
11. Gupta NK, Sekhon GS, Gupta PK (2005) Study of lateral compression of round metallic tubes. Thin Walled Struct 43:895–922
12. Morris E, Olabi AG, Hashmi MSJ (2007) Lateral crushing of circular and non-circular tube systems under quasi-static conditions. J Mater Process Tech 191:132–135
13. Celentano DJ, Chaboche JL (2007) Experimental and numerical characterization of damage evolution in steels. Int J Plast 23:1739–1762
14. Olabi AG, Morris E, Hashmi MSJ, et al. (2008) Optimised design of nested circular tube energy absorbers under lateral impact loading, Int J Mech Sci 50:104–116
15. Olabi AG, Morris E, Hashmi MSJ, Gilchrist MD (2008) Optimized design of nested oblong tube energy absorbers under lateral impact loading. Int J Impact Eng 35:10–26

16. Zeinoddini M, Harding JE, Parke GAR (2008) Axially pre-loadedsteel tubes subjected to lateral impacts (a numerical simulation). Int J Impact Eng 35:1267–1279
17. Zuraida A, Khalid AA, Ismail AF (2007) Performance of hybridfilament wound composite tubes subjected to quasi static indentation. Mater Des 28:71–77
18. Mcdevitti TJ, Simmonds JG (2003) Crush of an elastic-plastic ring between rigid plates with and without unloading. J Appl Mech 70:799–808
19. Lee KS, Kim SK, Yang IY (2008) The energy absorption control characteristics of Al thin-walled tube under quasi-static axial compression. J Mater Process Tech 201:445–449
20. Lipa S, Kotełko M (2013) Lateral impact of tubular structure—theoretical and experimental analysis. Part 1: investigation of single tube. J Theor Appl Mech 51(4):873–882
21. Baroutaji A, Morris E, Olabi AG (2014) Quasi-static response andmulti-objective crashworthiness optimization of oblong tube under lateral loading. Thin Walled Struct 82:262–277
22. Baroutaji A, Gilchrist MD, Smyth D, Olabi AG (2015) Crush analysis and multi-objective optimization design for circular tube under quasi-static lateral loading. Thin Walled Struct 86:121–131
23. Yin H, Wen G, Liu Z, Qing Q (2014) Crash worthiness optimization design for foam-filled multi-cell thin-walled structures. Thin Walled Struct 73:8–17
24. Zahiri-Hashemi R, Kheyroddin A, Shayanfar MA (2014) Effect of inelastic behavior on the code-based seismic lateral force pattern of buckling restrained braced frames. Arab J Sci Eng 39:8525–8536
25. Gantes CJ, Livanou MA, Avraam TP (2014) New insight into interaction of buckling modes with stable post-buckling response. Arab J Sci Eng 39:8559–8572
26. Xie S, Zhou H (2014) Impact characteristics of a composite energy absorbing bearing structure for railway vehicles. Comp, B 67:455–463
27. Yu TX, Xue P (2010) Engineering plastic mechanics. Higher Education Press, Beijing, China
28. Zheng M, Zhao Y, Teng H, Hu J, Yu L (2015) Elastic limit analysis for elliptical and circular tubes under lateral tension. Arab J Sci Eng 40:1727–1732
29. Timoshenko S (1955) Strength of materials, part 1, elementary theory and problems, 3rd edn. Stanford University, Stanford, USA
30. Alexander JM (1960) An approximate analysis of the collapse of thin cylindrical shells under axial load. Quart J Mech Appl Math 13:10–15
31. Zhang TG, Yu TX (1989) A note on a "velocity sensitive" energy-absorbing structure. Int J Impact Eng 8:43–51

Printed by Printforce, the Netherlands